Withdrawn
University of Waterloo

STRATEGIES FOR THE SEARCH FOR LIFE IN THE UNIVERSE

ASTROPHYSICS AND SPACE SCIENCE LIBRARY

A SERIES OF BOOKS ON THE RECENT DEVELOPMENTS
OF SPACE SCIENCE AND OF GENERAL GEOPHYSICS AND ASTROPHYSICS
PUBLISHED IN CONNECTION WITH THE JOURNAL
SPACE SCIENCE REVIEWS

Editorial Board

J. E. BLAMONT, *Laboratoire d'Aeronomie, Verrières, France*

R. L. F. BOYD, *University College, London, England*

L. GOLDBERG, *Kitt Peak National Observatory, Tucson, Ariz., U.S.A.*

C. DE JAGER, *University of Utrecht, The Netherlands*

Z. KOPAL, *University of Manchester, England*

G. H. LUDWIG, *NOAA, National Environmental Satellite Service, Suitland, Md., U.S.A.*

R. LÜST, *President Max-Planck-Gesellschaft zur Förderung der Wissenschaften, München, F.R.G.*

B. M. McCORMAC, *Lockheed Palo Alto Research Laboratory, Palo Alto, Calif., U.S.A.*

H. E. NEWELL, *Alexandria, Va., U.S.A.*

L. I. SEDOV, *Academy of Sciences of the U.S.S.R., Moscow, U.S.S.R.*

Z. ŠVESTKA, *University of Utrecht, The Netherlands*

VOLUME 83
PROCEEDINGS

STRATEGIES FOR
THE SEARCH FOR LIFE
IN THE UNIVERSE

A JOINT SESSION OF COMMISSIONS 16, 40, AND 44,
HELD IN MONTREAL, CANADA, DURING THE IAU GENERAL ASSEMBLY,
15 AND 16 AUGUST, 1979

Edited by

MICHAEL D. PAPAGIANNIS
Department of Astronomy, Boston University

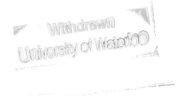
Withdrawn
University of Waterloo

D. REIDEL PUBLISHING COMPANY
DORDRECHT : HOLLAND / BOSTON : U.S.A.
LONDON : ENGLAND

Library of Congress Cataloging in Publication Data

DATA APPEARS ON SEPARATE CARD.

ISBN 90−277−1181−X
ISBN 90−277−1226−3 (pbk.)

Published by D. Reidel Publishing Company,
P.O. Box 17, 3300 AA Dordrecht, Holland.

Sold and distributed in the U.S.A. and Canada
by Kluwer Boston Inc.,
190 Old Derby Street, Hingham, MA 02043, U.S.A.

In all other countries, sold and distributed
by Kluwer Academic Publishers Group,
P.O. Box 322, 3300 AH Dordrecht, Holland.

D. Reidel Publishing Company is a member of the Kluwer Group.

All Rights Reserved
Copyright © 1980 by D. Reidel Publishing Company, Dordrecht, Holland
No part of the material protected by this copyright notice may be reproduced or
utilized in any form or by any means, electronic or mechanical
including photocopying, recording or by any informational storage and
retrieval system, without written permission from the copyright owner

Printed in The Netherlands

TABLE OF CONTENTS

FOREWORD

Leo Goldberg
Kitt Peak National Observatory
Tucson, Arizona 85726, U.S.A.

Of all the reasons for exploring the Universe, none is more compelling than the possibility of discovering intelligent life elsewhere in the Universe. Still the quest for extraterrestrial life has been near the bottom of the astronomers' list of priorities, not because the number of extraterrestrial civilizations is conjectured to be vanishingly small, but because our powers of detection were thought to be far too weak. About ten years ago, however, the growing reach of radio telescopes on the ground and of optical and infrared telescopes in space persuaded a number of thoughtful astronomers that the time for a more serious search had arrived. Accordingly, a joint Soviet-American conference on the problems of Communication with Extraterrestrial Intelligence was convened at the Byurakan Astrophysical Observatory of the Armenian Academy of Sciences during September 5-11, 1971 and was soon followed by a number of other important meetings, notably a series of NASA-sponsored workshops in the USA held between January, 1975 and May, 1976.

Since SETI is fundamentally an international undertaking and astronomical methods and techniques are required for its pursuit, it is natural for the International Astronomical Union to lend its support by sponsoring conferences and otherwise facilitating cooperation among countries. The active involvement of the I.A.U. in SETI was initiated by its sponsorship of the Joint Session on "Strategies for the Search for Life in the Universe" during the August, 1979, 17th General Assembly of the International Astronomical Union in Montreal. This meeting was jointly organized by three I.A.U. Commissions, #16 Physical Study of Planets and Satellites, #40 Radio Astronomy, and #44 Astronomy from Space. The preparation of the meeting was carried out by a 12-member International Organizing Committee of distinguished astronomers from the USA, USSR, U.K., France, Hungary, and Japan under the Chairmanship of Prof. Michael D. Papagiannis who is also the editor of this Volume of the Proceedings. To the Organizing Committee we want to express the appreciation of the astronomical community for their successful preparation of such a fine meeting.

M. D. Papagiannis (ed.), Strategies for the Search for Life in the Universe, vii–viii.
Copyright © 1980 by D. Reidel Publishing Company.

The four sections of the meeting, which comprise the four parts of this Volume, were attended by hundreds of astronomers from around the world. There was also an open evening session which I was asked to chair, and in which Drs. Drake and Papagiannis summarized the results of the Joint Session for the general membership of the I.A.U. This open session was held in the large auditorium of the University of Montreal and was attended by more than 1,000 astronomers and guests, with standing room only, characteristic of the interest generated in scientists and laymen alike by the subject of Life in the Universe.

This Volume of the Proceedings is sure to stimulate much new theoretical and experimental work on the part of the scientific community engaged in SETI. The enterprise will probably require several more decades of dedicated efforts before any meaningful results will be forthcoming. Most of the participants believe, however, that a significant program can be undertaken at modest cost and with substantial side benefits. The results obtained from such a long-term commitment by our scientific community will certainly be exciting even if we were finally to conclude that we on this planet are the only ones to have reached a high level of technology and cosmic awareness in the galaxy.

THE INTERNATIONAL ORGANIZING COMMITTEE

MICHAEL D. PAPAGIANNIS, Chairman   (U.S.A.)

FRANK D. DRAKE   (U.S.A.)

JUN JUGAKU   (JAPAN)

NIKOLAI S. KARDASHEV   (U.S.S.R.)

SIR BERNARD LOVELL   (ENGLAND)

GEORGE MARX   (HUNGARY)

TOBIAS OWEN   (U.S.A.)

MARTIN J. REES   (ENGLAND)

J.C. RIBES   (FRANCE)

I.S. SHKLOVSKII   (U.S.S.R.)

V.S. TROITSKII   (U.S.S.R.)

BENJAMIN M. ZUCKERMAN   (U.S.A.)

LIST OF PARTICIPANTS IN THE I.A.U. JOINT SESSION

AND OF CONTRIBUTORS TO THE VOLUME OF THE PROCEEDINGS

BAUM W.A.   Lowell Observatory, U.S.A.
BLACK D.C.   NASA Ames Research Center, U.S.A.
BOND A.   Culham Laboratory, England.
BEAKIRON L.   Van Vleck Observatory, Wesleyan University, U.S.A.
CURRIE D.G.   University of Maryland, U.S.A.
DRAKE F.D.   Cornell University, U.S.A.
GATEWOOD G.D.   Allegheny Observatory, Univ. of Pittsburgh, U.S.A.
GOEBEL R.   Allegheny Observatory, Univ. of Pittsburgh, U.S.A.
GOLDBERG L.   Kitt Peak National Observatory, U.S.A.
GREENSTEIN J.L.   California Institute of Technology, U.S.A.
GULKIS S.   Jet Propulsion Laboratory, U.S.A.
HART M.H.   Trinity University, U.S.A.
HOERNER S.V.   National Radio Astronomy Observatory, U.S.A.
JUGAKU J.   Tokyo Astronomical Observatory, Japan.
KARDASHEV N.S.   Space Res. Inst., Academy of Sciences, U.S.S.R.
KIPP S.   Allegheny Observatory, Univ. of Pittsburgh, U.S.A.
KUIPER T.B.H.   Jet Propulsion Laboratory, U.S.A.
MARTIN A.R.   Culham Laboratory, England.
MORRISON P.   Massachusetts Institute of Technology, U.S.A.
MCLEAN I.S.   University of Arizona, U.S.A.
OLIVER B.M.   Hewlett-Packard Company, U.S.A.
OLSEN E.T.   Jet Propulsion Laboratory, U.S.A.
OWEN T.   State University of New York/Stony Brook, U.S.A.
PAPAGIANNIS M.D.   Boston University, U.S.A.
REES M.J.   Inst. of Astronomy, Univ. of Cambridge, England.
RUSSELL T.   Allegheny Observatory, Univ. of Pittsburgh, U.S.A.
SERKOWSKI K.   University of Arizona, U.S.A.
SHKLOVSKII I.S.   Space Res. Inst., Academy of Sciences, U.S.S.R.
SOFFEN G.A.   NASA Headquarters, U.S.A.
STEIN J.   Allegheny Observatory, Univ. of Pittsburgh, U.S.A.
STURROCK P.A.   Stanford University, U.S.A.
SULLIVAN W.T. III   University of Washington, U.S.A.
TARTER J.   University of California, Berkeley, U.S.A.
TROITSKII V.S.   Radiophysical Research Institute, Gorki, U.S.S.R.
ZUCKERMAN B.   University of Maryland, U.S.A.

PREFACE

Michael D. Papagiannis
Department of Astronomy, Boston University
Boston, Massachusetts 02215, U.S.A.

It was a happy moment when, in the morning of August 15, 1979, I stepped on the podium in Montreal to open the Joint Session of the I.A.U. on "Strategies for the Search for Life in the Universe", and to welcome all the participants on behalf of the International Organizing Committee which had labored for nearly two years to prepare this meeting. It will be an even happier moment when the Proceedings of this Joint Session will appear in print after an additional year of letters, telegrams and telephone calls all over the globe. All these efforts, however, are easily compensated by the successful completion of this undertaking and the many new friendships with colleagues from all around the world that were started thanks to this meeting.

In my brief welcoming address in Montreal I said: "It is an important new step and a unique opportunity to hold a meeting on Life in the Universe during the General Assembly of the International Astronomical Union, when many of the best scientists in this field from around the world are gathered under the same roof. Our subject is a sensitive one that can easily lead to misunderstandings. It is our responsibility, therefore, to provide sensible leadership and to show to the world that we can proceed in a scientific manner, without exaggerations and without premature headlines toward the resolution of this fundamental question." I have repeated this statement here because I believe that dignity and international cooperation are of fundamental importance in the success of this effort.

I would like to thank my fellow members of the International Organizing Committee, Drs. F. Drake, T. Owen, B. Zuckerman of the USA, N. Kardashev, I. Shklovskii and V. Troitskii of the USSR, B. Lovell and M. Rees of the U.K., J. Ribes of France, G. Marx of Hungary and J. Jugaku of Japan for their help and collaboration in organizing this meeting. Also the Presidents of the three I.A.U. Commissions that sponsored this Joint Session: Dr. T. Owen, Commission 16 - Physical Study of Planets and Satellites; Dr. H. Van de Laan, Commission 40 - Radio Astronomy; and Dr. R.M. Bonnet, Commission 44 - Astronomy from Space, for their support and encouragement in the preparation of our Joint Session. Also

M. D. Papagiannis (ed.), Strategies for the Search for Life in the Universe, xiii–xiv.
Copyright © 1980 by D. Reidel Publishing Company.

I want to thank my distinguished colleagues who chaired or co-chaired the four sections of our meeting, Drs. I. Shklovskii and P. Morrison, section I; Drs. B. Oliver and J. Jugaku, section II; Drs. J. Greenstein and N. Kardashev, section III; and Drs. M. Rees and S. von Hoerner, section IV; and the four of them, Drs. Morrison, Oliver, Greenstein and von Hoerner who wrote the introductions to the corresponding parts of this Volume.

I want to express my deep appreciation to Dr. L. Goldberg, a past President of the I.A.U., who chaired the very successful open evening session and has written the Foreword for this Volume. Also to all the contributors of this Volume and to all the participants of the Montreal meeting. Finally, I want to thank Ms. Jackie Kiernan of the Astronomy Department of Boston University for her invaluable help in the preparation of this Volume, and express my gratitude to Mrs. N.M. Pols-v.d. Heijden of the D. Reidel Publishing Co. for her help and consideration in the publication of these Proceedings.

This Volume of the Proceedings is divided into four parts reflecting the four sections of the meeting in Montreal. Each part is dedicated to a particular topic of the general subject and consists of an introduction and several papers by many of the best known scientists in this field. In addition I have prepared an introduction to this Volume which presents the highlights of the meeting and allows the reader to obtain a quick overview of this Volume. Finally, though no formal drafting of conclusions and recommendations was attempted in Montreal, I have included a brief closing section in which I have tried to summarize the results of this Joint Session, as F. Drake and myself did in Montreal during the special evening session for the general membership of the I.A.U.. I hope that, though somewhat subjective, these conclusions will be of help to the readers of this Volume. It is the hope of all of us that this meeting will stimulate further work around the world in the planning of a sensible strategy for the search for life in the Universe.

Michael Papagiannis (center) welcoming the participants of the Joint

Session in Montreal. The co-chairmen of the first section, Philip

Morrison (left) and Josef Shklovskii (right), wave to the audience.

HIGHLIGHTS

STRATEGIES FOR THE SEARCH FOR LIFE IN THE UNIVERSE
HIGHLIGHTS OF THE PROCEEDINGS

Michael D. Papagiannis
Department of Astronomy, Boston University
Boston, Massachusetts 02215, U.S.A.

This introduction is intended to provide an overview of the Volume of the Proceedings of the Joint Session on "Strategies for the Search for Life in the Universe" that was held on August 15 and 16, 1979 in Montreal during the 17th General Assembly of the International Astronomical Union. The Joint Session was co-sponsored by three I.A.U. Commissions, #16 (Physical Study of Planets and Satellites), #40 (Radio Astronomy), and #44 (Astronomy from Space). The Volume is divided into four parts, as was also the Joint Session, that deal with four specific topics of the general subject. This introduction includes brief reviews of all the contributed papers, some comments from the discussions in Montreal, a few personal observations, and a short summary of the Soviet radio SETI program as I was able to put it together from various sources.

PART I.   THE NUMBER N OF ADVANCED CIVILIZATIONS IN OUR GALAXY
          AND THE QUESTION OF GALACTIC COLONIZATION.

Part I opens with an eloquent Introduction by Philip Morrison of M.I.T., who in 1959 wrote with Cocconi the classic paper suggesting a search for extraterrestrial civilizations at the 21 cm line of hydrogen. He parallels our different points of view on the value of N with the ideas of Aristotle, Malthus and Copernicus. He urges us to search the galaxy for "those swift, cheap manifestations of the existence of able, moderately-funded astronomers". We owe the issue more than theorizing, he writes, "because our Universe is still full of wonder and surprise".

The next paper N is Very Small is by Michael Hart, who initiated the discussion on galactic colonization with his so frequently quoted paper of 1975. He points out that if the number N of advanced technological civilizations in our galaxy was, as often quoted, in the 10^5-10^6 range, the entire galaxy including our own solar system would have been colonized a long time ago. The lack of any such evidence on earth convinces him that N must be very small, which he attributes

3

M. D. Papagiannis (ed.), Strategies for the Search for Life in the Universe, 3–12.
Copyright © 1980 by D. Reidel Publishing Company.

to a very low probability for the spontaneous origin of life on a planet. He dismisses the idea that in spite of the absence of any extraterrestrials N could still be large because "no intelligent extraterrestrials are interested in exploration and colonization" or because
"they do not want to interfere with us". The reason is that such
arguments assume a universal rule that must apply to all galactic civilizations, "a rule that is not followed by the only advanced civilization we know, the human species, which does travel, explore, colonize
and interfere with others".

The paper N is Neither Very Small Nor Very Large is by Frank
Drake of Cornell University, who formulated the famous "Drake equation"
to estimate the value of N, and who in 1960 with project Ozma carried
out the first radio search for extraterrestrial intelligence. Drake
favors a moderate value for N, of the order of 10^5, because interstellar travel and stellar colonization are too expensive and therefore
all galactic civilizations are bound to stay away from them. He estimates that if there was going to be interstellar travel, the whole
galaxy would be colonized in about 10 million years. He computes,
however, that the energy needed to send a colony of 100 people to another star would cover the present energy needs of the United States for
one hundred years. Consequently, no galactic civilization is likely
to turn to stellar colonization as a result of population pressures,
because the energy needed to send a few people to another star in
search of a better life, would provide many more with an affluent life
in their own solar system. He calculates, therefore, that the most
probable value of N must be a moderate one, because "the workings of
biology, the physical laws of energy and the vast interstellar distances conspire to make interstellar colonization unthinkable for all
time".

The paper Galactic-Scale Civilization is by Thomas Kuiper of the
Jet Propulsion Laboratory, who introduces some very interesting thermodynamic concepts with regards to the evolution of life. In principle
he believes, as discussed by Papagiannis in the next paper, that
"present knowledge favors the conclusion that N is either very large
or very small, but we do not yet have the means to distinguish between
these two possibilities". He proceeds, however, to make a good argument that N being very large is the more probable one because, as
first discussed by I. Prigogine, life consists of dissipative systems
that maintain or increase their internal order by dissipating order in
their environment. This is not a violation of the second law of thermodynamics, because the total entropy still increases, but suggests a
law of thermodynamics for life which Kuiper formulates as follows:
"An open system will eventually achieve the level of organization
(i.e., will decrease in entropy) necessary to maximize the rate of
dissipation of organization (free energy) in its environment". This
means that as life evolves it keeps consuming larger quantities of
energy to further increase its level of organization. The physical,
chemical, biological, social and cultural evolution represent clearly
a system of increasing organization. A technological civilization

seems to be an inevitable step in this sequence, which is likely to
lead to a galactic civilization as the next higher level of organization. "Our present knowledge on the absence of such a civilization",
he writes, "is only as good as the extent of our searches for it. An
eventual discovery that we are the first to reach this stage of development would be surprising."

The paper <u>The Number N of Galactic Civilizations Must be Either
Very Large or Very Small</u> is by Michael Papagiannis, who is also the
Editor of the Proceedings. He points out that modern ideas on space
colonies have eliminated the need for fast interstellar trips at almost relativistic velocities. Also the need for earth-like planets,
which might indeed be very rare in the galaxy, as a prerequisite for
stellar colonization. He postulates four principles about life:
I. Life tends to expand to occupy all available space; II. Adapts to
the requirements of every available space; III. Evolves continuously
toward higher levels of organization; and IV. The higher the level of
organization the faster the organization increases. Life, therefore,
behaves like an insatiable sponge for energy which it uses to grow and
conquer all available space. Barring, therefore, any insurmountable
difficulties, intelligent life is likely to follow the much faster
path of galactic colonization rather than wait for the painfully slow
evolution to high intelligence in the different solar systems of the
galaxy. He believes the reasons that will send man to the stars will
not be population pressures but rather "the spirit of exploration and
conquest, the challenge of lofty goals, the desire to be different and
the irresist ble call of virgin new worlds". Several of the potential
difficulties are discussed, including the feasibility of space colonies, economic and social problems, and the extremely long colonization
periods suggested by Newman and Sagan. None of these difficulties
seem to be of such universal value that would eliminate completely
interstellar travel and colonization for hundreds of millions of civilizations (as required by $N \simeq 10^5$) that must have appeared over billions of years in the galaxy. One is led, therefore, to the conclusion
that either the colonization has already occurred, in which case N
must be very large ($N \simeq 10^{10}-10^{11}$), or that the colonization has not
yet taken place because N is very small ($N \simeq 10^0-10^1$) and very few
advanced civilizations did appear in the long history of the galaxy.
Distinguishing between these two alternatives should not be very difficult for our modern technology. The possibility that the colonization
of the galaxy is now in progress is very low, because the colonization
time represents only about 0.1% of the age of the galaxy.

The paper <u>Uncertainties in Estimates of the Number of Extragalac-
tic Civilizations</u> is by Peter Sturrock of Stanford University, who has
made a statistical analysis of the values of N obtained by different
authors using the Drake equation. His analysis shows that the estima-
ted value of N lies in the 10^4-10^8 range with a 1σ ($\simeq 70\%$) confidence,
and in the 10^2-10^{10} range with a 2σ ($\simeq 95\%$) confidence. Remembering,
he writes, that "our estimate of the range is conservative, we see
that there is an enormous uncertainty in current estimates of N".

It should be noted here that the Drake equation assumes an independent evolution of life to high intelligence in each solar system and therefore does not include the possibility of galactic colonization which would tend to make the range of values for N even wider. Peter Sturrock also makes an attempt to estimate the time-span of the communicative phase of galactic civilizations, but this effort becomes quite esoteric because he tries also to incorporate the possibility that some of these civilizations might possess a knowledge of "hyperphysics" which none of us knows whether it even exists.

The paper A New Approach to the Number of Advanced Civilizations in the Galaxy is by V.S. Troitskii, who has been one of the leading Soviet scientists in this field. He introduces the idea of a simultaneous origin of life in the different parts of our galaxy, and possibly of the Universe. This represents a rather radical departure from the generally accepted concept of a continuous appearance of new centers of life in the galaxy, on which the Drake equation is based, but for which there is not as yet any experimental evidence. He proposes that life originated about 4 billion years ago on all the planets that at the time of this event had the necessary conditions for life. Such an impulsive origin of life, coupled with evolutionary periods to advanced civilizations considerably longer than 4 billion years, yields a small value for N and an absence of any old civilizations in the galaxy. This is in agreement with presently available evidence which suggests that the number of advanced civilizations in our galaxy might be very small.

PART II.   STRATEGIES FOR SETI THROUGH RADIO WAVES.

Part II opens with an Introduction by Bernard Oliver of the Hewlett-Packard Co., who is the originator of the well known project Cyclops. He states his firm belief that "radio is the only rational alternative for SETI" and explains why Kardashev type II civilizations and interstellar travel at 0.7c are unrealistic. Government funding of SETI radio projects, which he hopes will be initiated soon in several countries around the world, represents "a state of intellectual maturity which is most welcome for a society in which wars still exist".

The paper Microwave Searches in the U.S.A. and Canada is by Ben Zuckerman of the University of Maryland and Jill Tarter of the University of California, Berkeley and the Ames Research Center of NASA. Both authors are eminently qualified to write this review. Zuckerman, together with Palmer of the University of Chicago, conducted project Ozma II, probably the largest of all such searches. Tarter, on the other hand, is a key member of a team from Ames Research Center and the Jet Propulsion Laboratory that has carried out another one of these searches and is currently planning a far more ambitious search program which, with the use of mega-channel spectrum analyzers, should be able to increase the coverage of the search space (sensitivity, bandwidth coverage, frequency resolution, sky coverage, etc.) by several orders of magnitude. They give an excellent summary of the 6 most sensitive radio searches, out of a total of almost 20 that have taken place during the past 20 years in the United States and Canada. Most of the six major searches described in this paper were conducted at 21 cm, though the OH line at 18 cm and the H_2O line at 1.3 cm were also used in some of the searches. Most searches focused on several hundred of the nearby stars, but there was also an all sky search by the Ohio State group while Drake and Sagan looked into several of the nearby galaxies.

The paper A Bimodal Search Strategy for SETI is by Samuel Gulkis and Edward Olsen of the Jet Propulsion Laboratory and the California Institute of Technology, and Jill Tarter of Berkeley and Ames. The authors describe the strategy and give the technical details of a radio search that their group from JPL and ARC is planning. The objective of this project is to expand by several orders of magnitude the parameter space that has been covered up to now, using already existing radio telescopes. The radio search will be carried out in the microwave window of the earth (roughly from 600 MHz to 25 GHz) which encompasses the "water hole" and water line frequencies and where the radio brightness of the sky is minimal. The project will have two objectives and hence two modes of operation. One will be to survey the entire sky to a relatively constant flux level over a wide range of frequencies, which will ensure that all potential life sites will be observed at least up to certain flux level. The other will be to observe a number of preselected sites of greater promise over a small frequency range but with a much higher sensitivity. If properly funded, this project is scheduled to become operational by 1984.

We had also hoped to receive a review of the Radio Searches in the U.S.S.R., but unfortunately our Russian colleagues said very little on this subject in Montreal. This is a brief summary that I was able to put together from different sources. A summary of the Soviet SETI program appeared in Astron. Zh., 51, 1125, 1974 and in English translation in Soviet Astronomy, 18, 669, 1975. A search involving 10 of the nearest stars was conducted in 1968 at the Radiophysical Research Institute in Gorki, U.S.S.R. A much larger search program was carried out during the decade of the seventies as a collaborative effort of several observing stations around the Soviet Union including Ussuriysk, Kara-Day in the Crimea, Tuloma in the Murmansk region, Pustyn in the Gorki region and on some occasions the ship "Academician Kurchatov" cruising near the equator. The objective was to detect sporadic radio signals from outer space in the 3-60 cm range where the atmospheric and industrial noise is much lower. The wide separation of the different stations made it possible to gain a better understanding of the spacial origin of these signals. The results of this search were published by Troitskii, et al., in Acta Astronautica, 6, 81, 1979, where they report that a good number of signals were simultaneously recorded in several of these sites. They appear, however, to originate in the ionosphere or in the magnetosphere of the earth and, though they had not been previously observed, they seem to be of natural origin because

their intensity and rate of occurrence showed variations that followed
the 11-year cycle of solar activity.

In a note from V.S. Troitskii, which arrived as I was preparing
to mail the camera ready pages to the publisher, he reports that a new
project called "Obzor" is now being developed at the Radiophysical Research Institute (NIRFI) in Gorki, U.S.S.R. It will consist of a large
number of small, 1 m in diameter, parabolic antennas and will have a
wide angle coverage of the sky at 21 cm. A set of 25 antennas is
planned for 1981 to be increased to about 100 by 1985. A frequency
band of 2 MHz will be divided into 10 channels of 200 KHz each with an
anticipated sensitivity of 10^{-19} W/m^2. The search strategy will probably use some of the ideas on ETI convergence proposed by P.V. Makovetskii (Icarus, 41, 178, 1980), namely the use of natural beacons in
the galaxy such as the Crab Nebula and especially the appearance of
new novas, to help greatly reduce the space and time parameters of the
search space.

PART III.   THE SEARCH FOR PLANETS AND EARLY LIFE
            IN OTHER SOLAR SYSTEMS.

Part III opens with an Introduction by Jesse Greenstein of the
California Institute of Technology, who also chaired NASA's workshops
on planetary detection. He discusses the likelihood of stars having
planets, the difficulties of the present methods of detection and the
possibilities for the future. This is a topic of much broader interest
to astronomy, which after staying at the speculative stage for a long
time, has finally achieved respectability and content. "In a decade
or two we should hope to know whether solid platforms for life exist
elsewhere" and if such planets are identified, we can search with spectrometers for several compounds such as water vapor, oxygen and methane that can provide a strong evidence for the presence of life.

The paper The Astrometric Search for Neighboring Planetary Systems
is by George Gatewood and his colleagues at the Allegheny Observatory
of the University of Pittsburgh, who give a comprehensive review of
past, present and future capabilities of astrometric techniques. Astrometric searches of nearby stars for substellar companions have been
going on, mostly at the Sproul Observatory, for nearly thirty years.
Several reports have appeared in the literature announcing Jupiter-
like (similar mass and distance from their star) planets especially
around Barnard's star, but they have not been confirmed. So far only
companions with masses larger than 10 times the mass of Jupiter have
been confirmed. The difficulties in the interpretation of the data
stem basically from several inherent problems of the photographic methods employed. The Allegheny Observatory is currently engaged in a
study of 20 of the nearest stars with imporved photographic precision.
The accuracy so far of this study has been 5 mas (milliarcseconds) per
season but will improve further if the project is continued for 12
years as planned. For comparison, astronomers on Barnard's star, at a

distance of about 6 light years from our sun, will need an accuracy of 2.5 mas to detect Jupiter from the wobbling motion of the sun and an accuracy of 1.5 μas to detect the earth. The photographic plates will be replaced in the near future by photoelectric detectors which are two orders of magnitude more efficient and therefore will be able to detect Jupiter-like planets in nearly 100 of the nearest stars. A specially designed 81 cm refractor with a vacuum housing is under study to fully utilize the capabilities to about 0.2 mas/season of these detectors. Much higher accuracies can be obtained with spaceborne telescopes which eliminate the problems imposed by the earth's atmosphere. It is estimated that a 2.1 m telescope in space, equipped with an electronic multichannel astrometric photometer (MAP), will be able to reach accuracies of about 0.2 μas/year and thus will be able to detect earth-like planets around any of the nearly 500 stars within a distance of 500 light years from our sun.

The paper Search for Planets by Spectroscopic Methods is by Krzysztof Serkowski of the University of Arizona. He describes his technique which is using a high precision spectrograph to measure minute Doppler shifts in the radial velocities of stars that result from the orbiting of a star and a planet around their common barycenter. For the detection of Jupiter-like planets the method requires the determination of radial velocities to an accuracy of 5 m/s. Since the spectrograph operates in a 250 A range around 4,250 A, the required accuracy necessitates a spectral resolution of 0.00007 A, which Serkowski and his colleagues believe is within the capabilities of the present state of the art.

The paper The Search for Planets in Other Solar Systems through Use of the Space Telescope is by William Baum of the Lowell Observatory. The Space Telescope, he writes, which is expected to go in orbit during the 80's, can make an important contribution in the search for extrasolar systems through its wide-field/planetary camera system, better known on the CCD camera. This system with exposures of 1,000 seconds may be able to achieve astrometric precisions of 1-2 mas for stars down to the 22nd magnitude, which would allow us to search for Jupiter-like planets in at least 100 stars.

The paper A Comparison of Alternative Methods for Detecting Other Planetary Systems is by David Black of the Ames Research Center of NASA. It is a very illuminating review of the pros and cons of the different methods and of the overall prospects for the future. The direct detection of planets is extremely difficult because it is like trying to see a firefly sitting at the edge of a powerful search light. Still there is a slim possibility that the space telescope might be able to see some. Other direct methods that have been proposed include occulting disks, apodized images, and infrared interferometers in space. The infrared region is chosen because as in our case around 30 μ the sun is only 10^4 times brighter than Jupiter, while in the optical region it is 10^9 times brighter. Of the indirect methods, the astrometric technique seems to be the most promising one especially

from space telescopes, but it requires long dedicated efforts for both
humans and instruments. Spectroscopic and astrometric methods are
actually complementary, because the observable effects of the first
increase when a planet is located closer to the star, while of the second increase when a planet is located farther from the star. Another
indirect method is based on interferometric observations, either with
a single aperture (specle or amplitude interferometry) or with a multiple aperture system. The interferometric techniques were discussed
in Montreal also by Douglas Currie of the University of Maryland.
There have been many advances in planetary detection in recent years,
both in instrumentation and techniques. The rewards in the long run
will be many, not only for SETI but also in understanding what kind of
planets are formed at certain distances from stars of different types.
The difficulties are still many, but as Jesse Greenstein said once "We
are only limited by our willingness to invest time, thought and money".

The paper The Search for Early Forms of Life in Other Planetary
Systems is by Tobias Owen of the State University of New York at Stony
Brook, who looks into the future after the detection of earth-like planets in other solar systems. He presents several convincing arguments
in favor of life as we know it, i.e., life based on carbon and water
which might produce an oxygen atmosphere. He also discusses the evolution and present state of the different planets and major satellites
of our solar system drawing conclusions about the masses and distances
of planets that could maintain liquid water on their surface over long
periods of time. After the detection of such planets, spectrophotometric observations can reveal the presence in their atmospheres of water
vapor and/or of oxygen in concentrations that indicate the presence of
life. The simultaneous existence in the atmosphere of both oxygen and
methane constitutes a strong evidence for their biological origin. It
might even be possible to detect spectroscopically the presence of a
technological civilization from traces of artificially produced gases,
such as fluorocarbons, in their atmosphere. All these studies, of
course, are still in the future, but with the coming of space telescopes they might not be as far off as they seem.

In addition to all the astronomers in Montreal, we also had the
pleasure to have with us Gerald Soffen, a distinguished biologist
who had been in charge of the Viking missions to Mars and who is now
the Director of Life Sciences at NASA. His heavy administrative re-
sponsibilities did not allow him to prepare a paper for this Volume,
but his presence in Montreal was refreshing. It emphasized the multi-
faceted nature of the problem and the need for interdisciplinary collaboration. In spite of the fact that the Viking probes did not find
any life on Mars, it is essential to continue the exploration of our
solar system to find out how planets evolve, how stringent the requirements for life are, and how unique the earth might be. The discovery
in our solar system of any other form of life, Gerald Soffen noted,
would have completely changed our perspective of our search for life
in the Universe.

PART IV. MANIFESTATIONS OF ADVANCED COSMIC CIVILIZATIONS.

Part IV opens with a comprehensive Introduction by Sebastian von
Hoerner of the National Radio Observatory, who has been one of the
pioneers in this field. He begins with what he calls "the ergodic
theorem of technology" which states: "Whatever can be done with available matter, energy, and within the laws of physics, will be done given
enough time and space". He laments the colossal expenditures of all
nations on "defense" (about 400 billion dollars per year) and wonders
"why people intelligent enough to build nuclear bombs would really be
stupid enough to do so". He discusses space colonies, Bracewell
probes, astroengineering, Dyson spheres and radio beacons. Also project Daedalus and the colonization of the galaxy, which he estimates
can be accomplished in about five million years. On the cost-effectiveness of interstellar travel he points out that our nuclear stockpiles correspond now to nearly 100,000 Megatons of TNT, or 4×10^{27}
ergs, which could send 1,000 passengers to one of the nearby stars.
This of course would exhaust all of our nuclear bomb supplies, a statement that lit up the faces of all those present in Montreal because no
one could think of a better way to use up our nuclear stockpiles. He
thinks that the evolution toward a Galactic Club is a natural one,
noting that "what makes life worthwhile are the luxuries and great adventures that are not governed by cost-effectiveness and tend to go to
the limit of whatever is possible".

The paper Starships and Their Detectability is by Anthony Martin
and Alan Bond of the Culham Laboratory, England, two leading members
of the British Interplanetary Society. They give an extensive discrip-
tion of Project Daedalus which was prepared in the period from 1973 to
1978 by 13 scientists and engineers working under the auspices of the
B.I.S. This group tried to show that within the bounds of present
technology or of reasonable extrapolations to the end of the century,
an unmanned interstellar mission to Barnard's star is possible. The
propulsion will be with a nuclear pulse rocket where small spheres of
deuterium and helium-3 (instead of tritium to avoid a copious production of neutrons) will be ignited with high-power electron beams. The
total mass of the starship will be 54,000 tons, of which 50,000 tons
will be propellants. The cruising speed will be 0.12-0.13c, so that
the trip to Barnard's star will take only 50 years. One of the greatest problems of this design is the use of helium-3 which must be obtained from the atmosphere of Jupiter. They also review in this paper
the difficulties and the feasibility of different types of spaceships
and propulsion mechanisms that have been mentioned in the literature.
They divide them into two categories: small, fast starships with
V > 0.1c and large, slow "world ships" with V < 0.01c. The last ones
are conceived as "mobile homes" that could carry 250,000 people to the
stars in trips that would last many generations. They also discuss the
detectability of both types of starships. These must be very energetic
objects "releasing powers greater than 10^{18} watts on the average" which
should produce high levels of radiation all across the electromagnetic
spectrum.

The paper Radio Leakage and Eavesdropping is by Woodruff Sullivan III of the University of Washington. It advocates an intriguing idea namely that in addition to searches for purposeful signals we should also look for radio transmissions inadvertently leaking out from the planets of other technological civilizations. In our case, the most powerful radio signals escaping through the ionosphere of the earth are produced by military radars, which could be detected by an Arecibo-like radio telescope to a distance of 15 light years. Much richer in information, though about 100 times weaker in strength, are the narrow band ($\simeq 0.1$ Hz) video carriers of television signals which since different stations use different frequencies, form an excellent set of artificial spectral lines. From the Doppler shifts of these narrow lines, astronomers on other stars could determine the spinning rate of the earth as well as its orbital period. This would allow them to estimate the distance of the earth from the sun and hence the approximate temperature of our planet. It is interesting to note that over the entire electromagnetic spectrum, the earth outshines the sun, and by several orders of magnitude, only in these narrow frequency bands of man-made radio transmissions. Of course this is a very recent phenomenon and we have no way of knowing how long advanced civilizations may stay in this phase. It is possible, however, "that the unintentional leakage of powerful radio signals might continue much longer than a civilization would have the perseverance to transmit purposeful signals". It seems prudent, therefore, to keep both options open when searching for extraterrestrial radio signals.

CONCLUSIONS AND RECOMMENDATIONS.

A brief summary of conclusions and recommendations from this Joint Session was prepared by the Editor and is included at the end of this Volume. The key message that seems to emerge from the papers in this Volume and the discussions in Montreal is that a serious search for life in the Universe is warranted by the current state of our technology and the anticipated developments in the near future. A reasonable effort, well blended into our astronomical research and space exploration programs, could produce some truly impressive, whether positive or negative, results on the question of life in the Universe during the coming decades.

PART I

THE NUMBER N OF ADVANCED CIVILIZATIONS IN OUR GALAXY
AND THE QUESTION OF GALACTIC COLONIZATION

THE NUMBER N OF ADVANCED CIVILIZATIONS IN OUR GALAXY AND
THE QUESTION OF GALACTIC COLONIZATION.   AN INTRODUCTION

Philip Morrison
Department of Physics, M.I.T.
Cambridge, MA 02139, U.S.A.

Since the oldest of the digital computers, the fingers of the
two hands that as a rule are ten in number, we enjoy the marking of
twentieth anniversaries. It is now just twenty years since the
scientific literature began to inlcude papers dealing with the possibility of interstellar communication among putative societies spread
among the stars. The idea remains a speculation, but it has drawn
much attention, and with this symposium we find it reaching the level
of I.A.U. as a topic for daring discussion. If the idea should ever
prove correct, we may wish to modify the name - if not the acronym -
of this union to I.A.U., I for Interstellar!

Twenty years of experience is no petty guide to theory, though
it is not at all a reliable one. We see, for example, that the big-
gest radio dish of that day, Arecibo, then in the planning stages,
remains unmatched today for collecting area. We learn also that the
rate at which man-made objects spread into space is a rough mean
speed of some 1 or 2 A.U. per year, i.e., about $10^{-5}c$. Manned space
craft have moved slower by an order of magnitude and have reached out
only to a couple of light seconds over the whole twenty years. There
is no clear sight of strong tendencies toward acceleration, as yet,
though it is only fair to remind the reader that forecasts of human
activities are the least reliable of the theorist's predictions.

Space age or not, our means to reach into the galaxy and beyond
depends so far, as it has from the beginning, on our ability to handle photon streams. These travel at the speed of light and physically the galaxy is well-constructed to pass them freely over a very
wide band of frequencies to which the earth's atmosphere interposes
a familiar filter-passband. That set of channels does not demand so
much of science and technology as do material space probes; we astronomers have been increasing collecting areas and bandwidths and reducing noise figures and confusion limits over a long, long time.
The end is certainly not in sight.

M. D. Papagiannis (ed.), Strategies for the Search for Life in the Universe, 15-18.
Copyright © 1980 by D. Reidel Publishing Company.

All of this experience, and a growing number of more incisive studies have reinforced, at least for me, the view we took twenty years back: Let us search the Galaxy for those swift, cheap manifestations of the existence of able, moderately-funded astronomers, the photon streams, whether mere secondary leakage phenomena or the less sure but tantalizingly delectable deliberate beacons. This initial comment is no place to enter into the details of a search strategy; a whole session will try to summarize the experience and thoughts of those who have tried. Here my intent is only to offer a perspective.

Three or four undoubted facts lie behind any such enterprise. Let me list them:

i) Astronomers <u>can</u> exist for here we are. That we are here was dependent upon a star of spectral class G2 V is not to be ignored, even though we have no surety that such a circumstance is necessary. We do not know it to be sufficient either, for an unknown set of circumstances allowed the origin and evolution of copious life only on one planet among almost a dozen orbiting masses of some size. But it is hard to ignore the hint: let us look for some approach to symmetry with us out there too.

ii) Astronomers are a late phenomenon. If they are specified to handle radio frequency photons well, they may be assigned an age of some 10^{-8} Hubble time, or less. If one was to generously take their whole species into account, maybe you would increase the time by one or two orders of magnitude. Yet the system in which we dwell is about half a Hubble time old. It follows that we are truely jejune, real juveniles. Any other IAU's in the Galaxy have been founded long ago. Even the transit time is likely to be long compared to our own radio frequency history. The earth-sun system as a strong source of narrow-band microwaves is one of the newer phenomena in the Galaxy. Was there ever a predecessor? That is the issue.

iii) No journal has reported credibly any sign that out there are other astronomers, other engineers, other travelers in space. It is of course plain that we could not be seen either - although we are here - by any instruments we now have, save only the occassional flashing microwave beam of early-warning radar, Arecibo's planetary echo radar or perhaps the chorus of commercial TV carriers, none of which has yet crossed distances beyond our most immediate and provincial galactic neighborhood. That should give us caution: not one other earth is visible although our samples tell us that out there in the Galaxy there are a few times 10^8 sun-like stars, single stars of longevity, temperature, and stability close to those of our own Sun. But planets with radio probes would not yet be part of our catalog of galactic objects even if they were quite numerous and widely distributed, because, occasional pioneer searches we have made have sampled only a tiny fraction of the search space any planner would construct.

These points seem to me beyond doubt. The papers to come will try to carry this line of thought further: their aim is to offer and to support conjectures on the number N of communicative sources we might expect a priori out there among the stars (or even in other galaxies, for the truly imaginative).

The chief question which seems to arise in this now well-known framework is one of extrapolation in time. We can be sure that N, the number of potentially signalling sources, is at least 1 (or maybe a portion of unity for we are not yet fully trying!). We know by now that not every star or at least not every late F or early G dwarf is such a source, at least of narrow-band microwaves over a modest range of frequencies, for we have looked at a few stars pretty well, though certainly not conclusively. The fraction of stars examined is small; an estimate depends on detailed specification of beam angles, confusion limits and the like, though no one will put it more that a part in a thousand or so. But what will happen in the future? By analogy what has happened during their deep past, which remains still our own future? That is the tough problem for the speakers who follow.

It seems to me that we here mirror the thoughts of one or another philosopher of the past. I like to associate the name of famous Aristotle with the idea of a unique quality to this earth, and our life upon it. On such a point of view, we are never to find a signal. For the curious systems we call astronomers are known only here, and there is no indication at all that they can exist anywhere else in the universe. I cannot deny this; we have no adequate theories even for earth-like planets, let alone for the origin of life, its evolution over a very long time span, or the emergence of conscious, communicating beings with power to signal over interstellar intervals. I suppose some people belong to this category and would place the likely value of N externally as nil. The time extrapolation is hardly required.

A second point of view, much in evidence in this set of papers, is for me to be dubbed with the name of an English philosopher-critic of the Enlightenment, The Reverend Thomas Malthus. Malthus, who had a sharp eye for the exponential multiplication in numbers, was most influential, even upon Charles Darwin himself. I take the idea more generally: not only our population, but our power demands, our computing speed, our publication rate, and a good many other variables of science and society, show exponential trends. Now there are plenty of people today too who can draw straight lines on semi-log paper. They tend to infer that in no time we would very much increase our capability, not only of signalling, but even of physical travel, whether in person or by computer-probe proxy. Then, they argue, the evolutionary time is far greater than that required to spread a Malthusian growth of numbers and power right across the Galaxy! The absence of a sign of teeming Galactic technology must mean that we are first, or somehow unique. This argument is seductive, and it is surely correct IF the extrapolation of Malthus is justified in every necessary domain. IF!

No, I think I belong to the school of the philosopher Copernicus.
He said that we are ourselves a fair sample of what can be. The planets are like earths and the stars like other suns. Of course this
is not sure or even quantitative. It is only a point of view, a guide
to theory. Not all stars are sun-like, and Jupiter is far larger than
the earth. But I suspect that a society, a species, which finds
growth has limits, coming to a stop in many of the sub-systems from
local populations to energy use, is no poor guide to what goes on
elsewhere, though by no means at one and the same quantitative levels.
We will grow in power, surely, but not without limits, bounds and
reckoning of costs. I believe therefore we can look for creatures
better than ourselves, yes, but only within certain bounds. Not the
easy masters of whole stars or even galaxies, but rather of finite
domains, careful engineers, profligate like ourselves neither of
ships, energy or time. We shall see.

But I leave off with a good will to see what follows. It is
fine to argue about N. After the argument, though, I think there
remains one rock hard truth: whatever the theories, there is no easy
substitute for a real search out there, among the ray directions and
the wavebands, down into the noise. We owe the issue more than mere
theorizing. For, if any truth has entered the domain of the astronomer it is that once you look seriously in new ways for physical signals from the universe you often find something unexpected, even by
the most imaginative, because our universe is still full of wonder
and surprise.

N IS VERY SMALL

Michael H. Hart
Trinity University, San Antonio, Texas

ABSTRACT

If N, the number of advanced technological civilizations in a typical galaxy the size of our own, were a large number, then the solar system would probably have been colonized long ago. We infer that N is small. The most likely cause of N being small is that f_{life} (i.e., the fraction of suitable planets on which life actually arises) is very small; reasons are given why this should be so.

However, since the astronomical evidence indicates that the universe is open and infinite, there are an infinite number of inhabited planets, even though the mean number per galaxy is very low.

I)   THE DRAKE FORMULA

In calculating N, the number of advanced technological civilizations in a typical galaxy the size of our own, many persons use as a starting point some version of the formula suggested by Frank Drake in 1961:

$$N = R_* \cdot f_p \cdot n_e \cdot f_{life} \cdot f_{int} \cdot f_c \cdot L.$$

Here R_* stands for the rate of star formation; f_p for the fraction of stars with planets; n_e for the mean number of habitable planets per planetary system; f_{life} for the fraction of life-bearing planets on which life actually arises; f_{int} for the fraction of life-bearing planets on which intelligent species evolve; f_c for the fraction of intelligent societies which develop the ability to communicate with other worlds; and L for the mean longevity of such civilizations in the communicative state.

The various factors in the Drake equation are all highly uncertain, and estimates of them have varied widely, as can be seen from Table I. By choosing a combination of the more optimistic estimates, some authors have concluded that there are a billion or more advanced civilizations

M. D. Papagiannis (ed.), Strategies for the Search for Life in the Universe, 19–25.
Copyright © 1980 by D. Reidel Publishing Company.

TABLE I

SOME ESTIMATES OF THE FACTORS IN THE DRAKE-SAGAN EQUATION

	R_*	f_p	n_e	f_{life}	f_i	f_c	L
VERY OPTIMISTIC	100	1	5	1	1	1	10^9
OPTIMISTIC	50	1/2	1	1	1	1	10^6
MODERATE	(20)	1/5	1/10	(1/10)	1/2	(1/2)	(10^4)
PESSIMISTIC	10	(1/40)	(10^{-3})	10^{-20}	(1/10)	1/10	10^2
VERY PESSIMISTIC	1	10^{-8}	10^{-6}	$<10^{-1000}$	10^{-4}	10^{-2}	50

Note: The figures circled were selected for illustrative purposes only (see third paragraph of text). It is not meant to be suggested that the circled figures represent the most likely values of those factors.

in our galaxy alone. However, by combining the median estimates in Table I, we arrive at N ≈ 100; while choosing a moderately pessimistic combination of estimates, such as those circled in Table I, leads to the conclusion that only about one galaxy in forty contains even a single advanced civilization. Of course, if the very pessimistic estimates on the bottom line are combined, N turns out to be vanishingly small.

This range of results should not surprise us. There is no way in which we can take seven unreliable numbers, each of them uncertain by at least a factor of two -- several of them uncertain by many orders of magnitude -- multiply them together, and expect to get a reliable figure. The Drake equation, which can yield results ranging from $N \gtrsim 10^9$ down to $N \lesssim 10^{-1000}$, may be intellectually stimulating, but it is totally useless as a means of discovering the true value of N.[1]

II)   IMPLICATIONS OF FACT A

How then can we determine the value of N? I would like to suggest that N is best estimated not by looking at the factors which <u>cause</u> it to have a particular value, but rather by examining the observable <u>effects</u> which might result from a particular value of N.

For example, suppose that N were a large number, say 10^6, and that there were a million advanced technological civilizations in our galaxy. Then surely some of those extraterrestrial civilizations would have explored and colonized the galaxy, as we have explored and colonized the Earth. Since several estimates [2,3,4,5] indicate that it would only take a few million years to colonize almost the entire Milky Way galaxy -- a time interval which is very short compared with the age (≈ten billion years) of that system -- if N were large the solar system would have been visited and colonized a long time ago. If that were the case, we would see extraterrestrials here, now.

But the fact is that we observe no extraterrestrials on Earth today, nor anywhere else in the solar system (an important piece of data, which I have previously[2] referred to as "Fact A".) There is no sign that the solar system has ever been visited, and quite plainly we have not been colonized. The reasonable conclusion is that N is not equal to 10^6, nor to any other large number.

Some people dispute this line of reasoning, saying: "How do we know extraterrestrials will engage in interstellar travel and colonization? Maybe they prefer to spend their time, money, and effort on art, or in contemplating God, or in improving their material standard of living, or in constructing bigger and better bombs. Maybe they are just not interested in interstellar travel and communication."

Fact A might well be explained in that fashion if N were a small number. However, such an explanation is inadequate if it is assumed that N is large, for to explain the lack of extraterrestrial visitors and

colonists by hypothesizing, "they are not interested in exploration and
colonization," the hypothesis would have to hold for every extraterrestrial civilization, and at every stage in its history after it developed
the technological capability to engage in interstellar travel. If even
one extraterrestrial civilization decided to colonize, it (or its
daughter colonies) would soon spread through the galaxy.

Most of us will admit that we don't know much about the psychology
of intelligent extraterrestrial organisms. Those people who say, "there
are millions of advanced races out there, but none of them want to explore
and colonize," are implicitly claiming that they do know something about
alien psychology. They are in effect asserting: "All intelligent aliens
are alike in that they are uninterested in interstellar exploration and
colonization." I think that such an assertion, whether explicit or
implicit, is completely unjustified.

Here on Earth we see a bewildering diversity of attitudes and
psychologies between different individuals, societies and species. It
is completely implausible that, despite a million extraterrestrial
civilizations, there is less diversity in space than there is on Earth.
To explain Fact A by saying, "no intelligent extraterrestrials are
interested in colonization," or "they don't want to interfere with us,"
is to assume a universal rule, though a rule which is not followed by
the only advanced species we have observed: the human species, which
does travel, explore, colonize, and interfere with others.

III) WHY IS N SMALL? A POSSIBLE EXPLANATION

The foregoing reasoning, being of a general nature, is valid even
if we do not know which term or terms in the Drake equation causes N to
be low. I would like to suggest, however, that the most likely culprit
is f_{life}, the probability that life will spontaneously arise on a given
habitable planet.

On such a planet, by definition, the surface temperatures will be
moderate, liquid water will be abundant, and simple compounds of carbon,
hydrogen, oxygen, and nitrogen will be common. Numerous experiments [6]
show that in such an environment amino acids are formed readily, as are
many other compounds which are important precursors of life.

This is a promising start; but for life -- as we normally understand
the term -- to evolve there must also be some sort of genetic material
present. In terrestrial organisms the genetic material consists of DNA
molecules, which are long polymers each of whose monomeric components is
one of four compounds called deoxyribonucleotides. As is well known,
in a DNA molecule the order of the nucleotide residues is of vital
importance. Two different nucleic acid molecules, even if they are of
the same length, will not in general be biologically equivalent if their
nucleotide residues are not arranged in the same order.

   Most of the possible nucleic acid molecules are biologically useless
(sometimes harmful), or are useful only in conjuction with a set of other
genes. Let us assume, however, that there exists a particular DNA molecule -- let us call it "genesis DNA" -- which if introduced into primitive
proteinoidal material will not only replicate properly but will function
in some biologically useful fashion. In other words, we assume (optimis-
tically) that formation of a single molecule of genesis DNA in the proper
environment is sufficient to start the process of Darwenian evolution. [7]

   Let us assume (though no experiments to date show this) that not
only amino acids but also the nucleotides will be readily formed in the
conditions prevailing on a primitive Earth-like planet. Let us further
assume that conditions will be generally favorable for the polymerization
of nucleotides, and that uniform helicity of the resulting strands is
favored energetically. Let us also assume that genesis DNA is a relatively short gene, consisting of only 600 nucleotide residues in a strand,
and that some as yet unknown chemical effect favors the formation of
nucleic acid strands of just that length on a suitable planet. Under
these optimistic assumptions, a wide variety of DNA strands, each about
600 residues long, will be formed spontaneously by random processes on
a suitable planet. What is the probability that at least one of them
will have its nucleotide residues arranged in the same order as genesis
DNA?

   We know that there are $\leq 2 \times 10^{44}$ nitrogen atoms in the atmosphere
or near the surface of the Earth. Since a 600-residue strand of nucleic
acid contains over 2000 nitrogen atoms, there could have been at most
10^{41} DNA strands existing at any one time on the primitive Earth. (This
is obviously a very strong maximum.) If we assume that each strand could
split up and recombine with other fragments at a rate of three times a
second, then in a billion years (= 3×10^{16} seconds) a maximum of 10^{58}
different strands might have been formed. [8]

   On the other hand, the number of possible 600-residue long strands
of DNA is 4^{600}, or about 10^{360}. Apparently, the chance of forming any
particular one of them, even in a billion years, was negligibly small
($<10^{-300}$.) However, the situation is not quite that bad. It is known
that at some points in a DNA strand one nucleotide residue can be
replaced by another without altering the biological effect of the molecule. Suppose we assume that at 100 different points on the genesis DNA
strand either of two nucleotides will be equally effective, and that at
400 other positions any of the four nucleotide residues can be placed
without impairing the ability of the resulting strand to function as
genesis DNA. Under these very optimistic assumptions, the chance that
an arbitrarily chosen strand of DNA can function as genesis DNA improves
to 10^{-90}. Nevertheless, even in a billion years the chances of spontaneously forming even a single strand of genesis DNA are only $10^{58} \times 10^{-90}$
= 10^{-32}. [9]

   Furthermore, even if all the optimistic assumptions stated above
are correct, this figure of 10^{-32} would probably still be an overestimate.

For in existing organisms, DNA can only direct protein synthesis if certain other complicated molecules called transfer RNA are already present; and the synthesis takes place only on a complex particle called a ribosome.[10] In the absence of preexisting transfer RNA and a ribosome, even if a strand of genesis DNA were formed it would probably be unable to function.

IV)   WHAT ARE WE DOING HERE?

If the probability of life arising on a given planet is so very low -- at best 1 in 10^{32} -- how come there is any life in the universe? Since the number of habitable planets per galaxy is $\lesssim 10^{11}$, and since there are $\lesssim 10^{11}$ galaxies in the visible universe, it would appear that -- barring divine intervention -- the odds against life arising on even a single planet (Earth!) in the entire universe are staggeringly large.

This paradox is resolved, however, when we consider modern cosmological theory. In recent years, a strong consensus has developed in favor of a big-bang cosmology, and the bulk of the evidence points to an open, expanding universe.[11] Unless we assume a very unusual topology, such a universe will be infinite in extent, with an infinite number of galaxies and an infinite number of planets. In such an infinite universe, anything which has a finite probability (no matter how low) of occurring on a given planet must certainly occur on some planet: in fact it must occur on an infinite number of planets![12] (It should be pointed out that this infinite universe is not some strange idea of the present author, but rather represents the most orthodox current model of modern cosmologists.)

The total number of inhabited planets is therefore infinite (a conclusion which is almost inescapable if modern cosmological theories are correct); but the probability of life arising in any particular galaxy is extremely low. In theory, most intelligent races will see no other civilizations in their galaxy, nor indeed in the portion of the universe which they can observe. This, of course, is completely in accord with Fact A, and with all other observational evidence to date.

V)   CONCLUSION

Usually, when calculations based on existing theories are compatible with the observations, the conclusions are readily accepted. That so many people are reluctant to believe that N is low is perhaps primarily a manifestation of wishful thinking. Note that to obtain a high value for N it is absolutely necessary to have $f_{life} \gg 10^{-30}$. But this can only be the case if one assumes that there exists some as yet unknown non-random abiotic process which tends to arrange nucleotide residues in a biologically useful sequence. While one cannot prove that no such process exists, one should be hesitant to postulate unknown processes when the observed facts can be explained without them.

REFERENCES AND NOTES

1. Aside from the difficulty in estimating the individual factors
    involved, the Drake equation is open to the serious objection that
    it includes no term which takes into account the effects of coloni-
    zation. Though seemingly tautological, Drake's equation actually
    contains the hidden assumption that interstellar colonization is
    impossible, or at any rate never occurs. If applied to a case (such
    as the number of Earth's continents on which there exist technological
    civilizations) where colonization is definitely known to have occurred,
    the Drake equation yields a totally erroneous result.
2. Hart, M.H. (1975), "An explanation for the absence of extraterrestrials
    on Earth." *Q. Jl. Roy. Astron. Soc.* 16, pp. 128-135.
3. Jones, E.M. (1976), "Colonization of the galaxy." *Icarus* 28,
    pp 421-422.
4. Papagiannis, M.D. (1977), "Could we be the only advanced technologi-
    cal civilization in the galaxy?" *Astron. Contrib. Boston U., Series
    II, No. 61.*
5. Jones, E.M. "Discrete calculations of interstellar colonization."
    Submitted to *Icarus* (1979).
6. For a partial list, see references at the end of chapter 7 in *The
    Origins of Life on Earth* by Miller, S.L. and Orgel, L.E. (Prentice-
    Hall, Englewood Cliffs, N.J.,1974.)
7. The simplest organisms we know of that are capable of independent
    existence contain hundreds of genes. So the assumption that there
    exists a single gene - genesis DNA --- which by itself can produce a
    viable organism is extremely optimistic. If we make the more
    moderate assumption that under primitive Earth conditions an
    organism containing only one hundred genes would be independently
    viable, the low probabilities calculated in this section would,
    roughly, have to be raised to the 100th power.
8. For similar calculations see: Argyle, E. (1977), "Chance and the
    origin of life." *Origins of Life* 8, pp 287-298; and also Yockey,
    H.P. (1977), "A calculation of the probability of spontaneous
    biogenesis by information theory." *J. Theor. Biol.* 67, pp 377-398.
9. It is, of course, possible that other compounds may exist which can
    serve as genetic material; however, such compounds are likely to be
    just as large and complex as DNA is.
10. Miller, S.L. and Orgel, L.E. (1974), *The Origins of Life on Earth*,
    pp 74-76.
11. Gott, J.R., Gunn, J.E., Schramm, D.M., and Tinsley, B.M. (1974),
    "An unbound universe?" *Ap. J.* 194, pp 543-553.
12. Ellis, G.F.R. and Brundrit, G.B. (1979), "Life in the infinite
    universe." *Q. Jl. Roy. Astron. Soc.* 20, pp 37-41.

N IS NEITHER VERY SMALL NOR VERY LARGE

Frank D. Drake
National Astronomy and Ionosphere Center
Cornell University
Ithaca, New York, U.S.A.

ABSTRACT

A number of estimates of N are based on minimum assumptions, and lead to "moderate" values of about 10^5. The suggestion that N may be very much larger is not supported observationally. The suggestion that the lack of extraterrestrial colonists on the earth means that N is very small is examined in detail. It is shown that the interstellar colonization scenario is, in fact, not plausible because even one colony calls for expenditures of energy sufficient to support hundreds of millions of individuals for hundreds of years. The laws of physics, biology, and interstellar distances thus make interstellar colonization unreasonable, and moderate values of N continue to be most plausible.

INTRODUCTION

A straightforward application of our knowledge of stellar information, primitive biochemistry, and biology, and evolution leads to estimates of the value of N, the number of advanced civilizations in our galaxy. A way to quantify the value of N was given long ago (Drake, 1965), and a modern discussion of the most probable values of N has been given by Goldsmith and Owen (1980). The most probable values for N so obtained are of the order of 10^5, and may be termed the "moderate" values. These are the values we arrive at by applying Occam's Razor, utilizing, admittedly, quite limited knowledge. Being the values obtained with the minimum of assumptions, these values of N are being used to guide the planning of searches for extraterrestrial intelligence. Greatly different values for N which do not violate Occam's Razor might change greatly our approaches to detecting other civilizations. Thus proposals of quite disparate values of N must be considered quite seriously.

PROPOSAL THAT N IS VERY LARGE

Recently, Kuiper (see this volume) has hypothesized that N may be very much larger than the "moderate" values. He suggests that the

M. D. Papagiannis (ed.), Strategies for the Search for Life in the Universe, 27–34.
Copyright © 1980 by D. Reidel Publishing Company.

development of intelligence is perhaps as basic a law of the universe
as are the laws of say, electromagnetism and quantum mechanics. However, as Kuiper recognizes, there is no observational data to support
this. In fact, the absence of signs of blatant intelligent activity
makes a weak argument against his hypothesis. Although this absence
of evidence is not conclusive proof that his hypothesis is wrong, it
indicates that any proof of the validity of his hypothesis may well
have to await the detection of other civilizations. The "moderate"
hypothesis remains the simplest and most supportable.

THE SCENARIO OF GALACTIC COLONIZATION

On the other hand, Hart (1975) and later Jones (1978) have considered the same absence of blatant signs of intelligent activity and
proposed that it implies that we are the only technical civilization
in the Milky Way. They argue that when there is a sufficient degree
of technical sophistication, a degree considerably beyond our own,
that population pressure will lead to interstellar colonization. They
suggest that colonization will be extended indefinitely until an entire
galaxy is colonized. They calculate that this will take very much less
time than the cosmic time scale. Thus the first civilization to achieve
this ability will colonize every habitable planet in a galaxy before
another civilization has a chance. They then argue that the absence
of such extraterrestrial colonists on earth means that we are the first
highly technical civilization, and thus the only one so far. They propose that N is 1. Should this be true, then a search for other civilizations should be directed towards other galaxies rather than our own.

The interstellar colonization scenario does lead strongly to the
conclusion that the first technical civilization will overwhelm a galaxy, almost no matter what values   the key parameters obtain. To
see this, consider perhaps the simplest version of that scenario. I
will call this the "Coral Model" because in it the civilization "grows"
within the galaxy in a manner and geometry closely analogous to the
growth of an aquatic coral. In this simple model, it is assumed that
when a technical civilization reaches a certain population, N_c, it has
the resources, both intellectual and in gross planetary product, to
launch a viable colony of individuals, N_0, on an interstellar colonization flight. It launches one or more such colonies.

Each colony travels a distance D at a speed v_s to a nearby habitable planet. There it lands and develops a new technical civilization.
When that civilization's population reaches N_c, then it launches one
or more colonies as before. This continues until all habitable planets
are colonized. See Figure 1.

In such a model, the distance from the original civilization to
the most farflung colony grows linearly with time, on the average. The
important question is the velocity of the "frontier", v_f, of this civilization. This is obtained as follows:

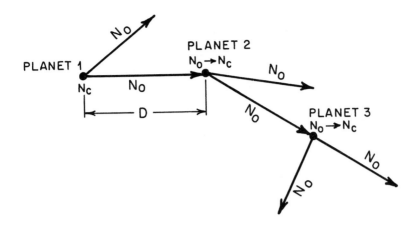

Figure 1.   The "Coral" Model of Interstellar
            Colonization

We assume exponential population growth since we expect that N_c
is far less than a population which strains a planet's natural resources.
Then

$$N_c = N_o^{\gamma t_c} \tag{1}$$

where γ is the fractional increase in population per unit time, and
t_c is the time interval required for a colony to reach N_c.

$$t_c = \frac{1}{\gamma} \ln \frac{N_c}{N_o} . \tag{2}$$

Let t_t = transit time of each colony, so

$$t_t = \frac{D}{v_s} . \tag{3}$$

Then the total time to establish each new colony is $t_c + t_t$, and v_f, the velocity of the civilization frontier, is

$$v_f \approx \frac{D}{\frac{1}{\gamma} \ln \frac{N_c}{N_o} + \frac{D}{v_s}} \qquad (4)$$

To see what v_f might be, it is interesting to take a plausible numerical example. Assume:

$$N_o = 1000$$

$$N_c = 10^8$$

$$D = 3 \text{ parsec}$$

$$v_s = 0.1 \text{ c}$$

$$\gamma = 0.01$$

Then $t_c = 1150 \text{ year}$

$$t_t = 98 \text{ years}$$

and $v_f \approx 2.4 \; (10^{-3}) \text{ parsec/year}$

With this value of v_f, a civilization would colonize an entire galaxy in a time of the order of 10^7 years, or only about one part in a thousand of the age of the galaxy. Thus in this scenario it is entirely to be expected that the first technical civilization will make its presence known on all inhabitable planets. The absence of extra-terrestrials on earth is then a puzzle, unless we are the first technical civilization.

    This conclusion is very insensitive to most of the assumed parameters. The transit time is only a small fraction of the total time per colonization step, so that an order of magnitude change in the velocity or more would have no qualitative effect on the result. The value of D might be as much as ten times larger, although even that is implausible, and this would have little qualitative effect. The values of N_0 and N_c hardly matter; N_0 could be 100, or N_c could be 10^9 and the value of t_c would change only by about 230 years, and v_f by only 18%.

SOME OBJECTIONS TO THE COLONIZATION SCENARIO

One possible explanation of the absence of extraterrestrials on earth is offered by the dependence of v_f on γ. A marked decrease in γ from the typical terrestrial value assumed here would increase the galactic colonization time almost in inverse proportion. Indeed, this suggests two possible explanations of the absence of extraterrestrials on earth.

First, if they adopt a policy of zero population growth, then there is no pressure to colonize in the first place, and the scenario never takes place. Or, secondly, if the population growth rate is very small, then the value of v_f may be so low that the galactic colonization time is greater than the present age of the galaxy. In this case, here and there in the galaxy there could be systems of interstellar colonies developing, but it is quite possible that none have reached earth yet. This would simply increase N over the "moderate" value somewhat, and would not alter the strategies for the detection of such civilizations.

Another explanation of the lack of terrestrial colonization has been offered by Newman and Sagan (1980), who suggest that each colonization step occurs in a randomly selected direction in space. In that case the frontier proceeds outward following the mathematics of a random walk, and it is easily shown that the galactic colonization time is more than the age of the universe. However, this seems an implausible scenario because it assumes that the extraterrestrials are unaware of the colonies which have preceded them, or where colonies already exist, and it seems certain they would have this information. The direction of their colonizations vectors would be preferentially radially outward if they were at all knowledgeable, and then the formulation given here is applicable.

THE LACK OF BENEFIT AS A MAJOR OBJECTION TO THE COLONIZATION SCENARIO

It can be argued strongly that the enterprise of interstellar colonization does not occur because of lack of benefit from it. In my opinion this is the most reasonable proposed explanation of the absence of extraterrestrials. One simply assumes that interstellar colonization will take place only if it offers sufficient benefits for the costs incurred. In fact, the distances between the stars and energetics of interstellar space flight seem to make that cost-benefit ratio very poor in all cases for interstellar colonization, implying that interstellar colonization does not occur.

To see why this is, it is very revealing to calculate how much
"space travel" can be purchased for an individual for the cost of
providing that same individual with a "good life" on the home planet.
In doing this, we can only use the present consumption of human beings,
of energy in particular, as a guide to what is required for a good life.
Of course, in more advanced civilizations the consumptions may be quite
different and can be both much greater or smaller, but as we will see,
this will probably have no qualitative effect on our conclusions.

With regard to possible space travel, we cannot of course know what
technologies may be employed. However, one thing we do know is that
the maximum kinetic energy of the space traveler and support facilities
will lead to a lower limit on the energy cost of any space colonization
venture. We can minimize speculation and assumptions about the future
of technology, then, if we simply adopt the energy now consumed to provide a human being with a "good life" as a basis to compute the maximum
kinetic energy of space travel for a colonist and support facilities
which would be considered cost-effective.

Thus, if we denote the total energy for a "good life" as E, and the
maximum velocity of a colonizing space craft as v_m, and the portion of
the mass of the spacecraft per colonist as M, then

$$v_m \leq \sqrt{\frac{2E}{M}} \tag{5}$$

Note that this leads to a high upper limit on v_m. As written, there
is an assumption that the propulsion system is 100% efficient, and
that no energy is required to decelerate the spacecraft at its destination. Also, no portion of the energy has been assigned to the production of the spacecraft and its fuel. It is also assumed that relativistic effects are negligible.

Now what are the actual values of E as given by the energy consumption of our civilization? We will use the United States as a basis
for this calculation, since it is the highest energy consuming country,
per capita, in the world. Then:

Total energy consumption of U.S.A. in 1979
(including consumption of oil, coal, hydro,
and nuclear energy resources) was..............10^{20} Joules

In 1979, population of U.S.A. was.......250 (10^6) people

Therefore, energy consumption was.......4 (10^{11}) Joules/
                                                      person-year
Now, assume average human lifetime is..........100 years;

Then, $E \approx 4 (10^{13})$ Joules

Now what of v_m? We have no good idea of the value of M, but it seems reasonable that M is not significantly less than 10 tons, which is slightly more than the mass per passenger of a large jet passenger aircraft. If we assume M = 10 tons, then

$$v_m = 90 \text{ Km/sec}$$

At first glance, this seems rather satisfactory, since it is a velocity some ten times greater than typical velocities of contemporary spacecraft. But, in fact, the result becomes entirely unsatisfactory when one calculates the time of transit. This time, for the 3 parsec flight which seems a reasonable minimum flight, is 40,000 years! The viability of a colony spacecraft and its inhabitants over a 40,000 year interval is very dubious, and such a journey would hardly be attractive to any potential colonists.

What energy is required for a much more reasonable transit time? Perhaps a 100 year transit time would be plausible, although many would argue that 100 years is still much too long. Assuming a 100 year transit time, the minimum energy per colonist increases by 400^2, and becomes:

$$\text{Minimum kinetic energy (100 years)} \approx 2(10^5) \, E$$

This same energy could, of course, provide a good life to about 200,000 people.

But the situation is much more energy demanding than this. It is much more realistic to include the inefficiencies in the production of fuel and in the spacecraft propulsion system. It is, of course, extremely difficult to estimate what these inefficiencies may be in an advanced technology, but it is probably not pessimistic to assume that the eventual kinetic energy of a space craft will be one-tenth of the energy expended in producing the fuel for the spacecraft. This still assumes no energy is required for landing. Then, in this more realistic case

$$\text{The Energy per colonist} \approx 2(10^6) \, E$$

A viable colony would require a rather large group of people of different talents; it is difficult to imagine such a colony containing less than 100 people. If we accept this estimate, then

$$\text{The Energy per colony (100 people)} \approx 200(10^6) \text{ E}$$

Now this is the total energy necessary to meet all the energy requirements of a major country, such as the U.S.A., for a period of time of
hundreds of years. It is very difficult to believe that any intelligent group of people, or any major government, would ever consider this
a cost-effective approach to any problem.

CONCLUSIONS

If, indeed, population growth creates a pressing need for lebensraum,
additional lebensraum can be created far more cheaply, as compared with
interstellar colonization, by pursuing such strategies as building
floating cities on the oceans or taking the heroic step of building
covered cities in the polar regions. Near-home space colonies of the
type advocated by O'Neill (1977) could always be built at a much smaller
cost per inhabitant than the cost of interstellar colonization. In
fact, it seems to me that this is the answer to the question "Where are
They?" and its rhetorical implication that "We may be the first." They,
even though there may be huge numbers of Them, are living comfortably
and well in the environs of their own star, thriving in habitats in
once uninhabitable climes or in the space surrounding the star.

In the end, it seems that the workings of biology, the physical
laws of energy, and the vast interstellar distances conspire to make
interstellar colonization economically unthinkable for all time. Our
best estimates of the numbers of civilizations then remain as they have
been, and the moderate values of N stand as the best estimates of N we
can make.

REFERENCES

Drake, F. D.: 1965, *Current Aspects of Exobiology*, ed. Mamikunian, G.
     and Briggs, M. H., Pergamon Press, Inc.

Goldsmith, D. and Owen, T.: 1980, *The Search for Life in the Universe*,
     The Benjamin/Cummings Publishing Co., Chapter 18.

Hart, M. H.: 1975, *Q. J. Roy. Astron. Soc.*, **16**, pp. 128.

Jones, E. M.: 1978, *J. Brit. Interplan. Soc.*, **31**, pp. 103.

Newman, W. I. and Sagan, C.: 1980, *Icarus*, in press.

O'Neill, G. K.: 1977, *The High Frontier - Human Colonies in Space*,
     William Morrow & Company, Inc.

GALACTIC-SCALE CIVILIZATION

T. B. H. Kuiper
Jet Propulsion Laboratory
California Institute of Technology

ABSTRACT

There is a remarkably consistent pattern of evolution in the Universe,
in which physical evolution leads to chemical evolution, followed by
biological evolution, social evolution, and cultural evolution. In
this process, matter becomes organized in increasingly complex ways,
and more able to make use of free energy in the environment. In view
of this evolution of dissipative structures, of which cells, humans,
and mankind are examples, it seems unlikely that technological civilization is an aberration in the Universe, but rather an inevitable step
in the process. Whether mankind is the first occurrence of this
evolutionary stage is not known, but the assertion that this is so is
unjustified. The subsequent level of organization is or will be a
galactic civilization.

1. INTRODUCTION

I must confess at the outset that I think that present knowledge
favours the conclusion that the number of sites of civilization in the
Galaxy (N) is either large or else it is small, and that we do not now
have the means to distinguish between these possibilities. I can only
hope to defend the proposition that N is large as a viable possibility
and perhaps even lead you to think that it might be the more probable
possibility.

2. THERMODYNAMICS

There is one argument that I cannot answer logically. That is the
proposition that we don't know anything at all, and so anything is
possible. To be able to discuss this topic, we have to admit that we
know something. When we look at evolution at its largest scale, we are
immediately aware that it is characterized by increasingly higher
levels of organization (Figure 1). This is why it has sometimes been

M. D. Papagiannis (ed.), Strategies for the Search for Life in the Universe, 35–43.
Copyright © 1980 by D. Reidel Publishing Company.

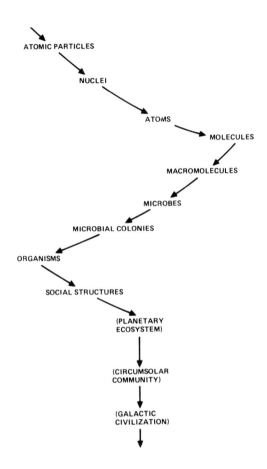

Figure 1 - Evolution presented as a cosmic process which results in
ever more complex organization of matter. The upper branch comprises
physical and chemical evolution, the next branch biological evolution,
and the third social evolution. Potential levels of high organization are shown in brackets.

said incorrectly that evolution violates the Second Law of Thermodynamics. The striking thing about this pattern is its cosmic nature.
We see physical and chemical evolution, biological evolution, and
social evolution in a continuous sequence. It does not suggest a
freak of nature, but a distinct process at work.

    The mechanisms of evolution have been studied over the last decade
or so by Prigogine and his co-workers, and they describe the organized
biological and social structures, as well as many physical and chemical
structures, as dissipative structures (Prigogine et al 1972). Such
structures maintain or increase their internal order by dissipating
order in their environment. Such structures are stable but fluctuate
about their stable point. Phrased another way, such structures exhibit

a normal organization but, either in time or in individual organization, also exhibit deviations from the norm. If there is a higher
level of stability (higher organization - lower entropy - higher rate
of dissipation), then in time one or more fluctuations will reach it.
We note that many transitions seem highly improbable so that some are
inclined to speak of singular events, or even miracles. But it is
in the nature of fluctuations far from the mean to be improbable. For
evolution to proceed, it is sufficient that the transitions be possible.
Then, given only enough time, it will eventually happen. If there is
then a suitable feedback mechanism so that the new structure is a
stable form, a new level of organization has been achieved. If the
structure is also autocatalytic and competes favourably with others in
the contest for resources in the environment, it will show an exponential growth which levels off as the carrying capacity of the environment is reached. In the realm of physical examples, which we understand
better because the organization is less complex, we accept as a fact
that (given the right conditions) a LASER never fails to work for want
of a spontaneously emitted photon, a magnetohydrodynamic instability
never fails to develop for want of a random fluctuation, and a pot of
water never fails to boil. This suggests a law of thermodynamics which
might be stated thus:

> "An open system will eventually achieve the level of
> organization (i.e. will decrease its entropy) necessary
> to maximize the rate of dissipation of organization
> (free energy) in its environment."

We thus view a single cell, whether independent or part of an organism,
as a dissipative structure. Equally, the organism, whether alone or as
a member of a hive or tribe, is a dissipative structure. Equally, the
entire terrestrial ecosphere is a dissipative structure, driven largely
(though not exclusively) by the Sun's radiation. That the ecosphere
has achieved stability is clearly not true.

3. CURRENT DIRECTIONS IN EVOLUTION

     It is fashionable to suggest that civilization has superceded
evolution, because there is no survival value or threat attached any
longer to deviations from the human norm. Indeed, individual evolution is probably over. This is analogous to the situation in our own
bodies. Deviations outside a certain range of the norm, however
advantageous they may be to individual cells, are not tolerated. The
cells are either isolated or rejected or, failing that, the organism
dies and the mutation with it. Only deviations which benefit the
entire organism are tolerated. Cellular evolution is superceded by
organistic evolution. Similarly, in our society, individual evolution
has been superceded by social evolution. Note that cells, in the one
case, and organisms in the other, will continue to evolve in the sense
of morphological change, but the "survival of the fittest" law no
longer operates on the individual, only on the aggregate. Indeed,
this phenomenon can be used to distinguish between a genuine structure
and a mere unorganized collection of individuals.

We have in recent decades become very aware that we must achieve
a global stability in short order if we are to survive. We must
achieve an organization which takes into account all the components of
this planet and their proper interrelationship. Ecology, conservation,
global economics, world politics, all these are but facets of the need
to achieve a fully integrated ecosystem on Earth. At the same time
we recognize the opportunity of replicating our terrestrial habitat
at other locations in our solar system (O'Niell 1974; Johnson and
Holbrow 1977). I do not think that there are many at this symposium
who would doubt that, if we achieve a planetary level of integration
(that is to say, if we do not destroy ourselves), that we will eventually establish ourselves elsewhere in the solar system.

The evolutionary step that concerns us here, however, is the next
one, the transition from a stellar civilization to a galactic one. I
think that the proper approach here is not to ask in detail how it
would be done, but rather to ask whether there are any known reasons
why it could not be done. Interstellar travel has been looked at from
a technical point of view and I think we must conclude for now that it
is, in principle, possible (Jackson and Whitmore 1978; Vulpetti 1979;
Matloff 1979; Martin and Bond 1979; Winterberg 1979). Then, given a
sufficiently large number of stellar civilizations, at least one will
be successful in colonizing another star. Once this is achieved, the
process will be repeated until the galaxy is entirely colonized. Note
that this does not imply that every star is inhabited. Our own Earth,
overcrowded as it is, has many isolated regions, and even human tribes
who have no contact with our globe-spanning civilization.

Various arguments have been presented against interstellar travel.
The proposition that it is impossible (e.g. von Hoerner 1962; Marx
1963) has been superceded by more sophisticated reasoning and largely
abandoned (e.g. von Hoerner 1978, 1979). It has been replaced by
social and economic arguments, i.e., that technological civilizations
choose not to travel between stars, perhaps because it represents too
great and economic burden (Drake 1979). An economic argument could be
compelling but it ignores the incredibly vast energy and material
resources in a planetary system such as ours. Economic analyses in
terms of current GNP's and similar indicators must be considered
meaningless. Social arguments are not compelling, because the great
diversity of possible social organizations and behavioural patterns
assures that at least some small percentage of capable civilizations
will actually engage in interstellar travel. Indeed, the evolution of
a nearby star towards a supernova is a powerful motivation for a star-
bound technological civilization to pull up its roots (de San 1978).

4.   PROBABILITIES

A number of simple calculations have been done to estimate the
time-scale of galactic colonization and, even with various assumptions
about travel speed and the time a colony requires to reach maturity,

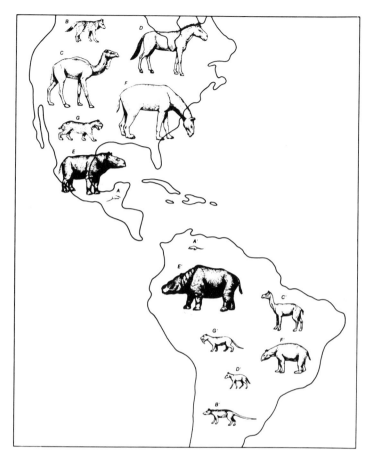

Figure 2 – Land mammals evolved independently in North America and
South America. Given ecological roles were played by phylogenetically
distinct animals on each continent (from May 1978).

the period of colonization turns out to be very short compared to the
age of the galaxy (Hart 1975; Viewing 1975; Jones 1976, 1978). Just
as an example, imagine an actual travel velocity of c/100 and a period
of 1000 y for the new colony to achieve full population before it
initiates its own colonization venture. Imagine that the typical
distance covered in each colonization venture is only 1 parsec. Then
the time for the colonization wave to travel 30 kpc is 30 million
years. This is a very short time compared to the age of the galaxy.
One could artificially extend this time estimate by, for instance,
postulating a civilization which maintains a low growth rate suitable
for populated worlds even on the newly colonized worlds of its frontier
(Cox 1976; Newman and Sagan 1980). This supposes a degree of social
homogeneity which seems incredible when one considers the relative
isolation of the member worlds (i.e. round-trip communication times of
a decade or more). In any case, a civilization which indulged in a

leisurely rate of colonization would be surpassed by a more vigorous one. Thus we reach the conclusion that it is very improbable that the galaxy is just now in its colonization phase.

One could in principle establish that N is very small by showing that the probability of life evolving to the point reached on earth is very small (e.g. Kuiper 1978b). Biologists have often pointed out how improbable it is that, for example, man should have evolved from reptiles. The concatenated probabilities of having all the right mutations is vanishingly small (e.g. Barloch 1977; references cited in Tipler 1979). The argument ignores, however, the evidence which suggests that Nature has many ways of achieving the same end result. The species in figure 2 are genetically distinct (May 1978). Those in South America were marsupials whereas those in the North were or are true mammals. Yet we see in both cases such morphologically distinct forms as the sabre-toothed cat, the hippopotamus, and the camel. One wonders then whether a sentient species arose exclusively because of the unique qualities of primates coupled with fortuitous mutations, or because Nature has an ecological niche for a sentient species. Indeed, analogies with simpler physical systems suggests the former. The characteristics of dissipative structures such as convection patterns, LASERS, magnetohydrodynamic wave patterns, concentration patterns in chemical solutions, all are determined exclusively by the appropriate physical laws and the boundary conditions. (Indeed, a system which were independent of environmental constraints would, by definition, be a closed system.) Flexible, efficient, and diversified data processing, when coupled with an effective means to manipulate the environment, seems to have such high survival value that life must inevitably lead to technological civilization. It therefore seems unlikely that Mankind is unique in the sense that the phenomenon has never happened before and will never happen again (i.e. a miracle). Conceivably, the concatenated probabilities could be such that Mankind is the first occurrence of such a phenomenon in the Galaxy but belief in the special status of Mankind does not have a history of successful prediction.

5.   ON FERMI'S PARADOX

This brings us then to the famous paradox attributed to Enrico Fermi, which asks "Where are they?" This is considered the big gun in the "N is small" arsenal. One possible answer has been given by John Ball (1973) and is known as the Zoo Hypothesis. The name speaks for itself.

Another possibility (e.g. Kuiper and Morris 1977; Kuiper 1978a) is indifference on the part of the galactic civilization. This would be especially likely if the independent evolution of sentient species were a relatively common event in the galaxy. Our own planet contains nearly the full range of human social evolution coexisting at the same time. Since we are being forced to stabilize our global civilization in a very short time, it is not clear that parts of the earth will not remain out of contact with our global civilization for an indefinite period into the future. We could well be one of the primitive tribes of the galaxy, not worth bothering with, or perhaps carefully protected.

Yet another possibility is that a star-faring civilization loses the need or desire to tie itself to individual stars (de San 1978; Stephenson 1979) but considers interstellar space its normal environment. Other possibilities are presented in the science fiction literature. The point is not that one of these explanations is necessarily correct, but that there are enough plausible explanations, including many we haven't yet imagined, that the "Fermi Paradox" cannot be considered a paradox at all.

6. SOME PHILOSOPHICAL REMARKS

It seems to be very difficult to remove anthropocentrism from our considerations of Mankind's status in the Universe. In the "Fermi Paradox," for example, lies buried the assumption that Mankind must be so interesting a phenomenon that a galactic civilization (think of a social organization with one planet for every person on Earth!) is compelled to establish contact with us, or alternatively, that the Earth is so desirable an environment that this civilization would have colonized it long before Mankind ever evolved. Since neither of these has happened, therefore Mankind must be a unique phenomenon in the Galaxy (if not the Universe)! Should this conclusion be inexplicably in error, then extraterrestrial beings must be so like Mankind (never mind the consequences of possibly several billions of years of further evolution) that we must surely be able to recognize their radio signals, or infrared waste heat, or other artifact! A far more humble attitude seems indicated.

One of the impediments to general recognition of intelligence as an integral part of cosmic evolution seems to be a belief that intelligence and civilization are transient phenomena. Surely individuals are born and die, societies rise and fall, and even whole worlds are undoubtedly destroyed in stellar cataclysms. In any case, eventually everything either cools asymptotically to absolute zero or collapses into a black hole. What significance then, can intelligence and civilization have on a cosmic scale? Such an attitude should be rejected out of hand as relying too much on our limited understanding of Nature.

Freeman Dyson has given one example to show the potential consequences of intelligent evolution (Dyson 1979). The example, which may be phrased in several different ways (e.g. Kuiper 1979), supposes that an intelligent entity consists of a finite number of quantum systems with two energy levels separated by energy ΔE. Intelligent activity consists of using energy from a finite reservoir to raise some systems from the lower to the upper level, and rejecting energy

into the cosmic void while lowering some other systems from the upper
state to the lower one. Let a "thought" be, on the average, m simultaneous operations of each kind. In order to minimize errors due to
spontaneous decays, or perhaps just due to enthusiasm, let the entity
function at the maximum rate allowed by the Uncertainty Principle, so
that a thought is performed in a time $\Delta t = h/\Delta E$. Let the entity think
n thoughts in a duration $T_1 = n\Delta t$ expending $nm\Delta E$ of energy from its
reservoir. Thereafter, let the entity restructure itself so that the
quantum levels are separated by $\Delta E/2$. To think n thoughts will now
take $T_2 = 2n\Delta t$ and consume $nm\Delta E/2$ of energy. Thereafter, it again halves
the separation of its energy levels. Since the series $E_1 + E_2 + E_3 \ldots$ con-
verges, the entity can clearly think an infinitude of thoughts with a
finite amount of energy in its reservoir, subject only to two constraints. It must have an infinite amount of time to think these
thoughts, and the cosmic radiation field must always be cold enough to
be a sink for the waste quanta (i.e. the Universe must expand forever).
There is a coherent body of evidence that suggests that the Universe is
open and will expand forever (Gott et al 1974). However, should our
Universe be rather different from what we now think, some other line of
reasoning may show the potential for intelligence to evolve to cosmic
scales, perhaps one day to determine the very nature of the Universe
itself.

7. SUMMARY

     In summary, a remarkably consistent pattern of development in the
Universe has been recognized for about a century (see Kuiper 1978b).
Work in non-equilibrium thermodynamics has gone some way towards
identifying the cause of this pattern. When fully understood, this
may well lead to a revision of the Second Law of Thermodynamics which
would state that, under the appropriate conditions, order increases as
inevitably as it decreases under conditions which we understand better.
If such a law indeed exists, then a phenomenon like mankind must inevitably arise and, given enough such occurrences, a galactic civilization will eventually come into being. Our knowledge about the present
absence of such a galactic civilization is only as good as the extent
of our searches for it. Our eventual discovery that we are the first
to reach this stage of development would be surprising.

     The research described in this paper was carried out at the Jet
Propulsion Laboratory, California Institute of Technology, under NASA
Contract NAS7-100. Permission to reproduce the figure from "The
Evolution of Ecological Systems" by Robert M. May (copyright ©️ 1978 by
Scientific American, Inc.; all rights reserved) is gratefully acknow-
ledged.

REFERENCES

Ball, J. A.: 1973, Icarus, 19, pp. 347-349.

Barloch, Lord Douglas: 1977, Q. Jl. R. Astr. Soc., 18, pp. 157-158.

Cox, L. J.:   1976, Q. Jl. R. Astr. Soc., 17, pp. 201-208.

de San, M. G.:   1978, "Hypothesis on the Origin of UFO's" (Editics, Bologna).

Drake, F. J.:   1979, this symposium.

Dyson, F. J.:   1979, Rev. Mod. Phys., 51, 447-460.

Gott, J. R., III, Gunn, J. E., Schramm, D. N., and Tinsley, B. M.:   1974 Astroph. J., 194, pp. 543-553.

Hart, M. H.:   1975, Q. Jl. R. Astr. Soc., 16, pp. 128-135.

Jackson, A. A., IV: 1978, J. Brit. Interpl. Soc., 31, pp. 335-337.

Johnson, R. and Holbrow, C.:   1977, eds., "Space Settlements - A Design Study" (NASA SP-413, Washington).

Jones, E. M.:   1976, Icarus, 28, pp. 421-422.

Jones, E. M.:   1978, J. Brit. Interpl. Soc., 31, pp. 103-107.

Kuiper, T. B. H.:   1978a, in "Our Extraterrestrial Heritage," Proc. of IAAA Symposium (IAAA, Los Angeles) pp. 25-31.

Kuiper, T. B. H.:   1978b, in "The Search for Absolute Values in a Changing World," Procs. of Sixth Int'l Conf. on the Unity of the Sciences (ICF Press, New York) pp. 905-917.

Kuiper, T. B. H.:   1979, Second Look, 2, #1, pp. 35-37.

Kuiper, T. B. H. and Morris, M.:   1977, Science, 196, pp. 616-621.

Martin, A. R., and Bond, A.:   1979, J. Brit. Interpl. Soc., 32, pp. 283-310.

Marx, G.:   1963, Astronaut. Acta, 9, pp. 131-139.

Matloff, G. L.:   1979, J. Brit. Interpl. Soc., 32, pp. 219-220.

May, R. M.:   1978, Scientific American, 239, #3, pp. 160-175.

Newman, W. I. and Sagan, C.:   1980, preprint.

O'Neill, G. K.:   1974, Physics Today, 27, #9, pp. 32-40.

Prigogine, I., Nicolis, G., and Babloyantz, A.:   1972, Physics Today, 25, #11, pp. 23-28 and #12, pp. 38-44.

Stephenson, D. G.:   1979, Q, Jl. R. Astr. Soc., 20, pp. 422-426.

Tipler, F. J.:   1979, preprint.

Viewing, D.:   1975, J. Brit. Interpl. Soc., 28, pp. 735-744.

von Hoerner, S.:   1962, Science, 137, p. 18.

von Hoerner, S.:   1978, Naturwissenschaften, 65, pp. 553-557.

von Hoerner, S.:   1979, this symposium.

Vulpetti, G.:   1979, J. Brit. Interpl. Soc., 32, pp. 209-214.

Winterberg, F.:   1979, J. Brit. Interpl. Soc., 32, pp. 403-409.

THE NUMBER N OF GALACTIC CIVILIZATIONS MUST BE
EITHER VERY LARGE OR VERY SMALL

Michael D. Papagiannis
Department of Astronomy, Boston University
Boston, Massachusetts 02215, U.S.A.

ABSTRACT   Life has the tendency to expand and adapt to occupy all
available space and to evolve toward structures of higher organization. Interstellar travelling, on the other hand, at V=0.01-0.10c
and the establishment of space colonies must be within the capabilities of technologically advanced civilizations. As a result, the
complete colonization of the galaxy in a relatively short period of
about 10 million years appears to be an unavoidable consequence of
life and technology. If the number of advanced civilizations currently present in our galaxy was a moderate one, i.e., $N=10^5-10^6$, over
the past several billion years there must have appeared in the galaxy
10^8-10^9 advanced civilizations. It seems therefore very unlikely
that all these hundreds of millions of civilizations over billions of
years were for some reason(s) prevented from following the natural
trends of life and technology. Consequently we are led to the conclusion that either the colonization of the galaxy has already taken
place, in which case N must be very large ($10^{10}-10^{11}$), or that the
colonization has not yet occurred because very few advanced civilizations capable of initiating it have appeared in the galaxy over its
past history, in which case N must be very small (10^0-10^1). The
probability that the galaxy is in the process of being colonized now
is very low ($\simeq0.1\%$), because the infusion of intelligent life into
the whole galaxy is bound to occur in about 1/1000th of the life of
the galaxy, i.e., in cosmic terms almost instantaneously.

1. HISTORIC BACKGROUND

Efforts to estimate the number N of technologically advanced
civilizations in our galaxy have been based for the past 20 years on
the famous Drake formula (Drake, 1965; Shklovsky and Sagan, 1966).
This equation consists of a series of factors which represent the
probabilities of a star being the right type to sustain life, having
planets in the habitable zone, of life originating on one of these
planets, evolving to high intellignece and finally to a civilization
with advanced technology (Papagiannis, 1978a; Goldsmith and Owen, 1980).

45

M. D. Papagiannis (ed.), Strategies for the Search for Life in the Universe, 45–57.
Copyright © 1980 by D. Reidel Publishing Company.

The Drake formula is based on the assumption that each galactic
civilization had an independent origin and that all of them have gone
through the slow biological evolution from primitive microorganisms
to societies with advanced technology. The reason for this key
assumption is that in the late 50's and early 60's, when the problem
of communication with other galactic civilizations was for the first
time considered in a serious manner, the prevailing scientific opinion
was that interstellar travelling was impossible and therefore each
galactic society had to have an independent start and evolution
around its own star.

Interstellar trips, and hence the colonization of unoccupied
solar systems, were excluded because of the belief that the travel
time could not exceed a reasonable fraction of a human life, i.e.,
no more than about 10 years. Since the distances between nearby
stars are of the order of 5 light years, for a 10 year trip the velocity of the spaceship would have to be close to one half of the speed
of light. The energetics, however, are such that according to our
present knowledge of the laws of physics it is practically impossible
to build a manned spaceship that can travel at $V \simeq 0.5c$.

A further complication was the belief that the members of the
mission, after exploring a nearby solar system, had either to return
back to the earth or they had to find an earth-like planet to establish a permanent settlement. Returning back to earth would make the
trip at least twice as long, while searching in different solar systems for an earth-like planet would probably make it even longer. The
probability of finding another planet where men could walk out from
their spaceship and breathe freely the air, something quite common
in most television stories, is actually quite low. The reason is that
a very similar atmosphere presupposes a planet with a very similar
mass, at a very similar distance from a very similar star being at a
very similar stage of a very similar biological evolution. All in all
the obstacles appeared to be so formidable, that the scientific community had decided to leave interstellar travelling to the science
fiction writers and to the cereal boxes and accepted the premise that
all galactic civilizations must have had independent origins.

2. CHANGES OF PERSPECTIVE

The continuous expansion of the space program in the 60's and
the 70's (men on the moon, landings of probes on Venus and Mars, flyby missions to Jupiter and Saturn, etc.), as well as the new ideas
about human colonies in space (O'Neill, 1977), have changed dramatic-
ally our perspective on interstellar travel. It now appears reasonable that in the next 50-100 years we will have explored the entire
solar system and we will have established permanent human settlements
in the interplanetary space. In the beginning these settlements will
probably be scientific missions and will consist of only a small num-

ber of people (10-100) stationed in space on a temporary basis. In
time, however, these settlements will expand in size and number of
inhabitants, they will develop their own industrial capabilities and
finally they will become independent entities with a permanent population and a life of their own.

From this stage on, the proliferation of space colonies is likely
to proceed at a rapid pace and we might even have space colonies with
specific industrial, ethnic or even philosophical identities. People
will be born and live their entire lives in these space colonies, re-
taining only a very tenuous sentimental link to the mother planet.
Many of these colonies will ultimately sever even this feeble link by
setting themselves on independent orbits around the sun, rather than
staying in orbit around the earth. These orbits could bring them
closer to the sun, if solar radiation continues to be their main
source of energy, or closer to the asteroid belt, an excellent source
of raw materials, if nuclear fusion has already solved their energy
problems. For the inhabitants of these settlements, the space colony
will be their whole life and therefore it will make little difference
whether their spaceship is on a circular orbit around the sun or on
straight flight to another star. In the final analysis our own
planet is a large spaceship that is travelling around the galaxy in
250 million years and nobody seems to mind this colossal interstellar
trip.

These new perspectives have freed our thinking from two basic
constraints of the past, namely the need to reach another star in no
more than about 10 years and therefore to travel at $V \simeq 0.5c$, and the
need to settle on earth-like planets. For a community with a happy
life in their space colony, an interstellar trip can last for several
generations and therefore there is no problem with relatively slow
velocities in the range $V = 0.01-0.10c$. The need also for habitable
planets is eliminated, because, even if earth-like planets were to be
found these people are likely to continue to live in space colonies
to which they have become accustomed for many generations. They will
only need raw materials to replenish their supplies and to build more
space colonies which they can easily obtain from the planets, moons,
comets and asteroids of practically every solar system around a
type I star.

In this manner, after an interstellar trip of a few centuries
and a build-up period of several centuries, the new stellar settlement
will have acquired a strong industrial capability to allow her to
send out new missions to other nearby stars. It should be noted that
the second venture will be much easier than the first one, since
these people would already know how to build powerful spaceships and
they would already possess the experience of interstellar travelling.
They would also probably have the benefit of new technological developments which occurred on the mother planet during their long trip
and which in a few years would be sent to them via radio or laser

links. Thus the colonization wave will advance from star to star throughout the galaxy with a speed of about one light year per century. A step, e.g., of 10 l.y. between two stars, would be bridged in about 1000 years, 2-3 centuries for the interstellar trip at V = 0.03-0.05c, and 7-8 centuries for the new settlement to build a strong industrial basis that would allow her to launch new missions to other stars.

With the colonization wave advancing in the galaxy at a speed of V = 1 l.y./century, the entire galaxy, which has a diameter of 100.000 l.y., will be colonized in less than 10 million years since the colonization will most likely start from somewhere inside rather that from the edge of the galaxy. It is important to note also that once this process is initiated, it is very difficult to stop it because the centers of colonization will multiply very rapidly. After a mere 10,000 years, e.g., there will be more than 10,000 stellar settlements from which the colonization process can continue. Finally it is of key importance to note that the time needed to colonize the entire galaxy (≃ 10 million years) is in cosmic terms a very short interval because it represents only about 1/1000th of the age of the galaxy (≃ 10 billion years). The conversion, therefore, of a galaxy from a desolate desert to a cosmic city teeming with intelligent life would represent an almost instantaneous event in the process of cosmic evolution.

3.   FOUR PRINCIPLES GOVERNING LIFE

From the myriads of species that have inhabited our planet and the billions of years of biological evolution we can deduce several general principles that characterize life as we know it, and could very well have universal value because life everywhere, as discussed by Dr. Kuiper in this Volume, must be a dissipative system that grows and increases in organization at the expense of energy available in its environment. The four principles that govern life are the following:

I. Life tends to expand, like a gas, to occupy all available space. A swimming pool without chlorine turns rapidly green from a profusion of algae, while a closed room is quickly covered with spiderwebs. As a matter of fact, spiders were the first insects to return to the islands of Bikini and Eniwetok after nuclear tests had completely erased every form of life from these small islands of the Pacific.

II. Life adapts to the requirements of every available space. Life has managed to grow lungs to move from the sea onto the land, and wings to fly in the air. Life can even adapt so as to take advantage of alternating phases of the same space as in the cases of day-time and night-time animals living in the same area, or of crops growing on the same land at different seasons.

III. <u>Life evolves continuously toward levels of higher organization</u>.
Procaryotic cells evolved into eucaryotic cells with well organized
nuclei. Unicellular organisms evolved into multicellular organisms
with cells organized into groups with specific functions. The organization of the brain has steadily increased to provide more memory and
higher intelligence, and finally members of a species form groups,
often with a clear division of labor, which represents an even higher
level of organization.

IV. <u>The higher the level of organization, the faster it increases</u>.
It took nearly 3 billion years of biological evolution to produce the
first multicellular organisms, but less than 60 million years from
the tiny lemurs to the australopithecus and only 3 million years from
the australopithecus to modern man. The organization of man into
cities and nations has been accomplished in a mere 10,000 years, while
modern technology with airplanes, radio communications, nuclear energy, computers, spaceships, etc., has blossomed in less than a century.

Man has followed faithfully and as a matter of fact has excelled
in each of these four principles. He has occupied practically every
corner of this planet, has conquered the seas with ships and submarines, the air with baloons and airplanes and has even ventured now
with his spaceships into outer space. He has organized cities of
millions of people and has established a fabulous system of global
communications. He has built a colossal industry that is using the
energy resources of nature to his great advantage, and a mushrooming
technology that is rapidly changing the quality of life on this
planet. A characteristic example of this rapid progress is the case
of long-distance, non-stop trips. One hundred years ago it would
take about one month to cross the Atlantic at a speed of about
2×10^2 cm/sec, while today we can travel to the moon in a couple of
days, i.e., 100 times farther and 1000 times faster at a speed of
about 2×10^5 cm/sec.

There is little doubt that man will continue to adhere to these
four principles and will proceed in time to also conquer the outer
space, the greatest frontier of them all. The conquest of our solar
system is already within our reach. The harnessing of nuclear fusion
will make interstellar flights possible at speeds V = 0.01-0.10c and
some day in the coming centuries, barring any global catastrophe from
the perils of modern civilization, we will embark on our first interstellar trip. Thus man will continue to advance the natural tendencies of life, namely to occupy the outer space, the largest space
available, and to reach at the level of a galactic community an even
higher level of organization. The infusion, therefore, of life into
the entire galaxy, is a goal in perfect harmony with the four principles of life outlined above.

Since the colonization of the entrie galaxy can be achieved in
about 10 million years, while the independent evolution of life into
a technological society in a solar system takes about 3.5 billion

years, i.e., 100-1000 times longer, it is obvious that the colonization of the galaxy represents a far more economical solution for
nature. This, by the way, was also the solution that was followed in
the case of man populating the earth. Early men, starting probably
from Africa, moved from place to place, from continent to continent,
from island to island to populate the entire earth. It would have
taken much longer for the independent evolution of man on every single
island of our planet, and man having already reached a substantial
level of intelligence was not willing to wait. By the same token,
it would take much longer for the independent development of advanced
civilizations in the billions of stars of the galaxy, instead of
having one of the first advanced civilizations populate the entire
galaxy. Barring therefore any insurmountable difficulties, high
intelligence is likely to follow this faster path.

There are, however, people who believe that there are indeed
several serious difficulties that could prevent interstellar travelling and hence the colonization of the galaxy. We will examine some
of the most often mentioned difficulties starting with the feasibility
of human colonies in space, which represents the first step in the
colonization process. Their feasibility has been questioned by
Ed. Purcell of Harvard University (private communication).

4. HUMAN COLONIES IN SPACE

A space colony represents essentially an air bottle in the vacuum
of outer space. For a spherical shell of radius R, thickness D and
internal gas pressure P, the stress S experienced by the walls will
be,

$$S = \frac{P\pi R^2 - (-\pi R^2 P)}{2\pi RD} = P\frac{R}{D} \tag{1}$$

If the tensile strength of the shell is S_0 and the safety factor
allowed is f, then we also have,

$$S = f S_0 \tag{2}$$

and hence the thickness of the shell must be

$$D = \frac{PR}{fS_0} \tag{3}$$

If the density of the material is ρ, the total mass M of the shell
will be,

$$M = 4\pi R^2 D\rho = \frac{4\pi R^3 \rho P}{fS_0} \tag{4}$$

For an aluminum shell 1 km in diameter with S_0 = 30,000 p.s.i.
(2040 Atm/cm^2), ρ = 2.7 gr/cm^2, P = 1 Atm and f = 0.5 we obtain
D $\simeq$ 50 cm and M $\simeq$ 4.1 × 10^6 tons, which represents about 30% of the
current annual world production in aluminum.

Thus for a space colony with a population density q (inhabitants/unit surface area), which O'Neil (1977) estimates at about 10,000 people/km^2 including areas for agriculture, the shell mass m per inhabitant will be

$$m = \frac{M}{q4\pi R^2} = \frac{PR\rho}{qfS_o} \tag{5}$$

which for the example discussed above results to about 130 tons of aluminum per inhabitant. With a current aluminum price of about $1500/ton, the cost of the aluminum alone would come to about $200,000/inhabitant without including any costs for sending this mass into space from the earth or the moon, construction expenses in space, etc., which would probably increase the final cost per inhabitant by at least a factor of 10. The obvious question then is why spend all this money to give to 30,000 a supposedly better life in space, while with the same amount of money you would provide them with an even better life on earth.

Well, first of all a design consisting of n smaller spheres with the same total area $(4\pi R^2 = n4\pi R_n^2)$ and hence capable of accomodating the same number of people, would reduce m and hence the cost per inhabitant by a factor of $n^{\frac{1}{2}}$

$$\frac{m_n}{m} = \frac{R_n}{R} = n^{\frac{1}{2}} \tag{6}$$

Thus it would be 10 times less expensive to house 30,000 in 100 smaller spherical shells 100 m in diameter each, rather than to keep all of them together in a large spherical shell 1 km in diameter. These smaller units would be interconnected and arranged so as to allow for future expansion. A symmetric design with all these properties might be to arrange these sub-units in the form of a double helix, as shown in Figure 1. The advantage of such a design, in addition to a great saving in material costs, is the fact that it allows for a gradual construction and occupancy and therefore would not tax the world economy as a single gigantic project that is of no use until it is fully completed.

History, with innumerable examples (submarines, airplanes, radio and T.V., computers, man on the moon, etc.) teaches us that the dreams of yesterday become the hopes of today and the realities of tomorrow, provided of course that they do not violate any physical laws. Our industrial power is increasing rapidly and what seems difficult today might very well be quite simple in a century or two. One century ago there was hardly any aluminum available since the industrial production of aluminum from the electrolysis of purified alumina started only in 1886. The world production of aluminum is doubling approximately every 10 years. If this trend was to continue during the next 100 years, an aluminum mass which today represents 30% of the annual world production, in 100 years would represent only an insignificant fraction of 0.0003 of the world production.

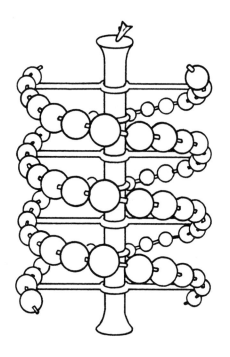

Figure 1. A Space Colony
with a Double-Helix design,
which provides for staged
construction and occupancy,
structural symmetry, and
needs far less material to
build it that an equivalent
large single unit.

Two centuries ago with the lilliputian world production in iron
(about 70,000 tons in 1780) the construction of an ocean liner such
as the Queen Elizabeth would have taxed the resources of the entire
world to an unacceptable limit. Today, however, including aircraft
carriers, supertankers, etc., there are probably more than 100 such
mammouth ships crossing the oceans, while the world production of
steel has increased to about 700,000,000 tons.

In conclusion, alternative construction designs, such as the
module approach shown in Figure 1, which would allow a substantial
reduction in cost and a staged construction and occupancy, as well as
our rapidly expanding industrial capability tend to minimize the magnitude of the economical problems as they appear today. It would be
a mistake, therefore, to believe that economic difficulties which
today might seem colossal will prevent man in the long run from establishing colonies in space.

5.   OTHER DIFFICULTIES

Several other problems regarding interstellar travelling and the
colonization of the galaxy have been discussed in the literature.
Let us consider the most important of them.

I.   The energetics of interstellar travel make such missions
essentially impossible. This problem was discussed in the past by

Ed. Purcell and is reiterated in this Volume by B. Oliver. The
statement of course is true if we are looking for velocities of about
0.5c or higher. There is no basic problem, however, with speeds in
the 0.01-0.05c range where nuclear fusion (Hart, 1975; Papagiannis,
1978b) could easily provide the energy for an interstellar trip without requiring an unmanageable load of fuel. The lower speeds, of
course, imply interstellar trips of several generations, which however, as discussed in Section 2, can be undertaken by independent
space colonies without any serious change in their daily routine.

     II. Emigration to other solar systems is totally uneconomical
and therefore it is not likely to occur as a response to population
pressures. This difficulty is discussed by F. Drake in this Volume.
This statement is also true if "population pressure" is thought of as
the only cause for interstellar travelling and stellar colonization.
The obvious reason is that emigration to space colonies, let alone to
other solar systems, will never solve the over-population problems of
a planet, because as we have seen in Section 4, it costs more to send
a person to a space colony than to provide him with an affluent life
on earth. It is also clear that sending a few hundred or a few
thousand people to other stars will never be undertaken to provide
more lebensraum for the inhabitants of the earth left behind.

     It could be undertaken, however, for other reasons such as the
reasons that convinced the United States to spend 10 billion dollars
in the 60's to send a handful of men to the moon, or make people risk
their lives to climb the Himalayans or to dive into the abysses of
the oceans. Interstellar travel will never be undertaken for practical purposes, though some benefits for mankind will undoubtedly result
from it as they did from the space program. It will be the indomita-
ble spirit of exploration and conquest, the challenge of lofty goals
and the need to excel, the desire to change and to be different, and
finally the irresistable call of virgin new worlds that will drive
men to the stars.

     It is true that the expenses of interstellar travel appear today
to be excessive (20% of the G.N.P. of the U.S.A. as mentioned by
F. Drake in his talk), though not totally beyond the means of our
civilization. The relative significance of this cost, however, will
certainly keep decreasing as the industrial capability of our society
increases in the years to come. It is very difficult, therefore,
to accept that what appears today to us to be too expensive, will
remain also prohibitively expensive for each and every one of the
millions of technological civilizations in the galaxy many of which
will have reached industrial and technological levels far superior to
ours. Interstellar travel, therefore, is bound to be undertaken by
some of these civilizations not as a means to solve their population
problems but as a natural response to the four principles of life
discussed above, an inevitable step to push life toward its ultimate
limits.

III.   Interstellar travel might be prohibited by social reasons. One can think of many good reasons (economic, religious, philosophical, political, etc.) that might prevent interstellar travel for some of the galactic civilizations during some period of their existence. It would be very difficult, however, as also pointed out by M. Hart and T. Kuiper in this Volume, to come up with a universal reason that would stop trillions of generations in hundreds of millions of different galactic civilizations from attempting interstellar travel. The assumption that such a reason could exist implies an exceptional level of social homogeneity in the Cosmos which is contradicted by the diversity observed on earth between individuals, nations, and historical periods. It is totally unreasonable, given the immense differences in the origin and evolution of the different galactic civilizations, to expect to find on a galactic scale a commonality and constancy of attitudes that we can not find among groups of people even in our own little world, inspite of our common origin. It is inevitable, therefore, that some civilization at some level of its development will initiate interstellar travelling and from that time on the complete colonization of the galaxy will be accomplished in a very short, in cosmic terms, interval.

IV.   Interstellar travelling is practiced but the complete colonization of the galaxy will take many billions of years. This problem has been advanced by Newman and Sagan (1980) based on the computations of the colonization time carried out using diffusion theory. Jones (1978,1980), on the other hand, using the Monte Carlo method, as well as Papagiannis (1978a,b), Hart (1975), Kuiper and Morris (1977), and Drake in this Volume, using simpler methods, have concluded that the colonization of the galaxy can be accomplished in about 10 million years. The complicated equations used by Newman and Sagan, as well as by Jones, tend to obscure the causes of this discrepancy by including immigration rates, random processes, diffusion rates, etc. The key factor, however, in all these calculations remains the time a new stellar colony needs before it will be ready to send out new missions to other neighboring stars.

Newman and Sagan (1980) have come up with waiting periods of the order of 10^5 years, which explains the very slow pace of galactic colonization. The process is slowed down even further because they assume a random walk expansion, which as pointed out by F. Drake in this Volume is not justified since these people will clearly know what stars have already been colonized in their vicinity and therefore will move only toward those that are still unoccupied. The exceedingly long waiting times postulated by Newman and Sagan are totally unrealistic given the fact that just about three centuries after the Pilgrims landed in America their decendants managed to send manned missions to the moon after developing the necessary technology from step one, while these stellar colonies are only asked to repeat a task that they already know well since this is the way they too established their own colony.

Newman and Sagan obtain their long waiting periods by imposing a very low rate of population growth ($\simeq 10^{-4}$/year) and by assuming that a relatively high, critical level of population must be reached before new missions to other stars can be undertaken. The population criterion, however, is the wrong one to use because, as discussed earlier, interstellar travelling will never be undertaken as a result of population pressures. The criterion ought to be the industrial preparedness of a civilization, which with the automation that characterized technological civilizations will undoubtedly advance much faster than the population growth. It is quite certain that the build-up of their industrial complex will proceed much faster than the few centuries it took on earth, since they would already possess the know-how and they would have probably drawn detailed plans for their industrialization even before arriving at their destination.

Thus a new stellar colony of about 10^4 people with a strong industrial basis, a level which can be achieved in less than 1000 years, would be ready to send out 300 of its members, i.e., about 3% of its population, to continue the process of galactic colonization. It should also be pointed out that the speed of colonization will be determined by the faster civilizations because the slower ones will be deprived of the opportunity since all of their nearby stars will have been colonized already by the fastest of their neighbors. There will certainly be a competition among existing galactic civilizations for new, unconquered worlds and therefore the slower civilizations will be automatically eliminated from the competition. Based, therefore, on the expected fast industrialization of advanced societies and the competitive nature of the colonization processes, one can rule out the extremely long waiting periods suggested by Newman and Sagan.

6. CONCLUSIONS

From the discussion above it follows that none of the difficulties suggested by people (economic, technical, or social) are likely to prevent interstellar travel or to slow down the colonization of the galaxy substantially beyond the 10 million year level. It is also interesting to note that even the proponents of the different difficulties (Purcell, Oliver, Drake, Sagan, etc.) do not agree among themselves what problem will ultimately prevent the colonization of the galaxy. As a matter of fact in most cases the majority of the opponents of the colonization disagrees with the obstacle proposed by one of them. Purcell, e.g., disavows the feasibility of space colonies. Sagan and Drake accept them. Oliver belives that interstellar travelling is realistic only at speeds around 0.5c and hence it is not feasible. Sagan and Drake accept interstellar travelling as speeds in the 0.01-0.10c range and therefore consider it feasible. Drake thinks that interstellar travelling and colonization will not be practiced because it is totally uneconomical as a means to alleviate population pressures. He accepts the idea, however, that if it is initiated it will proceed at a relative fast pace completing the colonization of the entire galaxy in about 10 million years. Sagan, on the other hand, has no argument with the economic difficulties presented by Drake but thinks, contrary to Drake, that the colonization process will be extremely slow competing in time-length with the age of the galaxy.

These disagreements on the causes that might prevent the colonization of the galaxy simply suggest that though each of the difficulties proposed might be applicable for some of the galactic civilizations at some stage of their development, none of them is strong enough or universal enough to apply to all galactic civilizations all of the time. It is inconceivable to have presently 10^5-10^6 advanced civilizations in our galaxy, and therefore to have had 10^8-10^9 civilizations (Freeman and Lampton, 1975; Papagainnis 1978a) in our galaxy over the past several billion years, and still to have avoided the complete colonization of our galaxy. Consequently the moderate value of $N = 10^5$-10^6 is unrealistic because either the colonization of the galaxy has already occurred and therefore N must be very large, i.e., $N = 10^{10}$-10^{11}, with space colonies in orbit around practically every well behaved star of the galaxy, or the colonization has not yet occurred simply because there have been very few, if any, other advanced civilizations in the past history of the galaxy to initiate the colonization process. This implies N must be very small, i.e., $N = 10^0$-10^1 and we could very well be the only technological civilization in our galaxy.

The conclusion, therefore, is that either the colonization of the galaxy has already occurred and N is very large, or that the colonization has not yet taken place because N is very small. The possibility that the colonization of the galaxy is now in progress and that only parts of the galaxy have so far been colonized is exceedingly small ($\simeq 0.1\%$) because the colonization of the galaxy is likely to be accomplished almost instantaneously in cosmic terms, i.e., in about 1/1000th of the age of the galaxy.

6. SEARCH STRATEGY

It should not be very difficult to find out experimentally whether N is very large or very small, because if N is very large there ought to be space colonies in orbit around all the well behaved stars of our galaxy, including our own solar system and several of the nearby stars. A careful radio, and possibly infrared and laser search of say the 100 nearest stars, together with a thorough exploration of our own solar system and especially of the asteroid belt (Papagiannis, 1978b), would allow us to decide if the galaxy has already been colonized. This is a manageable task that can be accomplished in a few decades as part of our overall astronomical research and space exploration program.

If we could make, therefore, a serious commitment to such a

program, it is reasonable to expect that in a few decades, and possibly by the year 2001 as prophesized by Arthur Clarke,we will know if our galaxy has been colonized or not. If it has, we will probably be invited to join as junior partners an already blossoming galactic supercivilization. If not, we will have to conclude that we must be one of the very few, if not the only advanced civilization in our galaxy. Such a conclusion would probably be a sobering experience, but also a unique challenge for our civilization to be given the opportunity to be the one to infuse intelligent life into the entire galaxy.

Both of these alternatives, N very large or N very small, represent fascinating prospects, and deciding which is the right one has been brought to within reach by our advances in science and technology. Standing at the threshold of such a cosmic insight, a mystery on which philosophers and scientists have speculated for thousands of years, makes our times a unique period in the history of mankind.

REFERENCES

Drake, F.D.: 1965, Current Aspects of Exobiology, ed. Mamikunian, G. and Briggs, M.H., Pergamon Press, Inc.

Freeman, J. and Lampton, M.: 1975, Icarus, $\underline{25}$, 368.

Goldsmith, D. and Owen, T.: 1980, The Search for Life in the Universe, The Benjamin/Cummings Publ. Co.

Hart, M.H.: 1975, Q.Jl. R. Astr. Soc. $\underline{16}$, 128.

Jones, E.M.: 1978, J. Brit. Interpln. Soc., $\underline{31}$, 103.

Jones, E.M.: 1980, Submitted to Icarus

Kuiper, T.B.H. and Morris, M.: 1977, Science, $\underline{196}$, 616.

Newman, W.J. and Sagan C.: 1980, Icarus, in press.

O'Neill, G.K.: 1977, The High Frontier-Human Colonies in Space, William Morrow and Co., New York.

Papagiannis, M.D.: 1978a, Origin of Life,ed. H. Noda, Center Acad. Publ., Tokyo, Japan, p. 583.

Papagiannis, M.D.: 1978b, Q. Jl. R. Astr. Soc. $\underline{19}$, 277.

Shklovskii, J.S. and Sagan, C.: 1966, Intelligent Life in the Universe, Holden Day, San Francisco.

UNCERTAINTY IN ESTIMATES OF THE NUMBER OF EXTRATERRESTRIAL CIVILIZATIONS

Peter A. Sturrock
Stanford University

ABSTRACT

Estimation of the number N of communicative civilizations by means of Drake's formula involves the combination of several quantities, each of which is to some extent uncertain. The uncertainty in any quantity may be represented by a probability distribution function, even if that quantity is itself a probability. The uncertainty of current estimates of N is derived principally from uncertainty in estimates of the lifetime of advanced civilizations. It is argued that this is due primarily to uncertainty concerning the existence of a "Galactic Federation" which is in turn contingent upon uncertainty about whether the limitations of present-day physics are absolute or (in the event that there exists a yet-undiscovered "hyperphysics") transient. It is further argued that it is advantageous to consider explicitly these underlying assumptions in order to compare the probable numbers of civilizations operating radio beacons, permitting radio leakage, dispatching probes for radio surveillance or dispatching vehicles for manned surveillance.

> The Master said, Yu, shall I tell you what knowledge
> is? When you know a thing, to know that you know it,
> and when you do not know a thing, to recognize that
> you do not know it. This is knowledge.
> Analects of Confucius (Waley's translation).

1. REPRESENTATION OF UNCERTAINTY BY DISTRIBUTION FUNCTIONS

The preceding contributions to this chapter have been concerned with various estimates of N, which denotes the number of extant advanced technical civilizations in our galaxy possessing both the interest and capability for interstellar communication (Shklovskii and Sagan, 1966). This number is estimated on the basis of a formula first presented by F.D. Drake when participating in a conference held at the National Radio Observatory at Green Bank, West Virginia, in 1961 (Shklovskii and Sagan, 1966, pp. 409 et seq.):

$$N = R f_p P_e f_\ell f_i f_c L. \tag{1}$$

M.D. Papagiannis (ed.), Strategies for the Search for life in the Universe, 59–72.
Copyright © 1980 by D. Reidel Publishing Company.

In this expression, R (year^{-1}) is the mean rate of star formation; f_p is the fraction of stars with   planetary systems; p_e is the mean number of planets in each planetary system with environments favorable for the origin of life; f_ℓ is the fraction of such favorable planets on which life does develop; f_i is the fraction of such inhabited planets on which intelligent life with manipulative abilities arises; f_c is the fraction of planets populated by intelligent beings on which an advanced civilization arises; and L (year) is the lifetime of the technical civilization possessing both the interest and capability for interstellar communication. (For reasons which will become clear later, I use p_e in place of the usual term n_e.)

It is standard procedure that, in presenting experimental or observational results, a scientist clearly indicates the uncertainty in his estimates, usually by the simple procedure of ascribing a standard deviation to the estimate. It is usually implicitly assumed that the distribution of the estimates is "normal" but of course this may not in fact be the case. In the absence of a statement concerning the form of the distribution (normal or otherwise), a statement of the standard deviation gives only a fragmentary representation of the estimated error.

It is not customary to present similar estimates of the uncertainty of theoretical calculations. Nevertheless, there is always uncertainty, if only because the theoretical model may or may not be a fair representation of the real physical system under consideration. The importance of this concept is discussed further in an article dealing with the evaluation of astrophysical hypotheses (Sturrock, 1973).

For the problem in hand, our knowledge of N is due to calculations, made for instance by means of the Drake formula (equation 1). Since there have been discussions in this chapter of the possibility that "N is very small" or "N is very large," etc., it is clear that there is in fact considerable uncertainty in our knowledge of N. It is therefore worthwhile to consider the sources of this uncertainty, how these sources contribute to the final uncertainty, and if possible to make some estimate of the final uncertainty.

If one were to take an estimate of N at face value, it would seem appropriate to adopt $N^{1/2}$ as a measure of the uncertainty, in accordance with the usual formulas of Poisson statistics (e.g. Wall, 1979). Since a typical estimate of N is 10^6 (Shklovskii and Sagan, 1966), this would imply that the accuracy of our estimate is 0.1%, whereas—as we have already seen—the uncertainty is very much greater. This fact underlines the need for a formalism which will lead to a more realistic estimate of the uncertainty. I suggest that an appropriate generalization of the Drake formula is one which replaces an estimate of each quantity by a distribution for that quantity.

It will be somewhat more convenient to work in terms of logarithms of the various quantities. We shall therefore write

$$\nu = \log N, \ \rho = \log R, \ \phi_p = \log f_p, \ \text{etc.}, \ \tilde{\omega}_e = \log p_e, \ \lambda = \log L. \quad (2)$$

We now characterize our assessment of a quantity x by the distribution P(x), such that P(x')dx' is the probability, according to one's analysis of specified information, that the quantity x lies in the range x' to x' + dx'.

Noting that, in terms of our new variables, equation (1) becomes

$$\nu = \rho + \phi_p + \tilde{\omega}_e + \phi_\ell + \phi_i + \phi_c + \lambda, \quad (3)$$

we see that our generalization of the Drake equation becomes

$$P(\nu) \quad = \quad \int \ldots \int d\rho \ldots d\lambda \ P(\rho) \ldots P(\lambda) \delta(\nu - \rho - \ldots - \lambda). \quad (4)$$

A distribution function P(x) gives much more information about an assessment of a quantity than is given by a simple estimate of that quantity. Although one may present an assessment of the uncertainty by means of the width of the distribution function (as measured, for instance, by the standard deviation), the distribution-function representation is a more flexible way of characterizing uncertainty. For instance, we may note that Sagan (Shklovskii and Sagan, 1966), in his discussion of L, considers two possibilities: an "optimistic" one that L ~ 10^9, to which he ascribes a probability of order .01; and the alternative (probability 0.99) "pessimistic" possibility that L ~ 10^2. He forms from these lifetimes the mean lifetime $\langle L \rangle$ ~ 10^7 and uses this in the Drake formula, as is quite appropriate if one is simply trying to determine the expectation value of N. If, however, one is interested in estimating the probability that N falls in some range of values (for instance, N in a range of very small values, such as would follow from the short lifetime L ~ 10^2), it is preferable that one retain the two alternatives explicitly. It is furthermore desirable that each of the two possibilities (optimistic and pessimistic) should be represented by a distribution function. Even if one pursues only one chain of argument to estimate a quantity, there will normally be a certain range of uncertainty about the estimate which should be represented by a distribution function.

In scientific work, one is continually relying on information supplied by one's colleagues. For instance, the best estimates of the seven quantities R, f_p, etc., might be obtained from seven different specialists. However, it often happens that one obtains information about the same quantity from two or more sources. If, as is likely, these sources do not agree, one then has the problem of somehow combining these estimates. If each of a large number of scientists makes a simple estimate of the quantity, then one effectively obtains a distribution of that quantity which one may be able to represent simply by a mean value plus a standard deviation. Suppose, however, that estimates are obtained from a small number of sources--say two. Suppose also that one source has a great deal more information and experience than the other. How then should one combine the two different estimates?

The use of distribution functions offers a possible answer to this
question. Each scientist can represent his estimate by a distribution
function. If one scientist is (presumably for good reasons) very sure
of his estimate, his distribution function may be quite sharp. If the
other scientist is uncertain of his estimate, his distribution function
will be broad. Following arguments given elsewhere (Sturrock, 1973),
we can combine two (or more) estimates $P_1(x)$ and $P_2(x)$ for the same
quantity x as follows:

$$P_{12}(x) = \frac{P_1(x) \, P_2(x)}{\int dx' P_1(x') \, P_2(x')} . \tag{5}$$

It normally happens that the confidence which a scientist places in
each of his sources differs (perhaps substantially) from the confidence
which each source places in himself. In this case, the scientist would
not accept the distribution functions $P_1(x)$, $P_2(x)$, at face value. As
a guide to a possible procedure to use in this case, we may note that if
two independent estimation procedures were to lead to the same function
$P_1(x)$, the resulting estimate is represented (to within a normalization
factor) by $[P_1(x)]^2$. This suggests the generalization that, if it is
necessary to "weight" an estimate, this may be done by replacing $P_1(x)$
by $[P_1(x)]^{w_1}$.

This rule would be particularly helpful if two sources give estimates which are so different and so sharp that they are irreconcilable.
It would then be prudent to replace P_1 and P_2 by P_1^w and P_2^w where w is
made sufficiently small that the resulting functions have a reasonable
chance of representing the same quantity--i.e., they have a healthy amount
of overlap.

In order to present a brief numerical discussion, we now assume
that each distribution function on the right-hand side of equation (4)
has a gaussian form. Strictly speaking, this assumption cannot be
correct since f_p, f_ℓ, f_i and f_c lie in the range zero to unity, and p_e
takes non-negative integer values. Nevertheless, by virtue of the
Central Limit Theorem (Wall, 1979), it is likely that our final distribution for N will be insensitive to this deficiency in our assumptions,
and in fact that it will be approximately gaussian. With this simplify-
ing assumption, the mean values of the quantities are related by

$$\overline{\nu} = \overline{\rho} + \ldots + \overline{\lambda}, \tag{6}$$

and the standard deviations are related by

$$\sigma^2(\nu) = \sigma^2(\rho) + \ldots + \sigma^2(\lambda). \tag{7}$$

As is clear from earlier contributions to this chapter, estimates
of N vary considerably due to variations in the estimates of the quantities R, f_p, etc. A few estimates of these quantities have been gathered
together in Table 1. Based on this information alone, we may obtain

estimates of $\bar{\rho}$, $\sigma(\rho)$, etc. These estimates are given in the last comumn
of Table 1. However, one should regard these estimates of the uncer-
tainty of each quantity as being <u>underestimates</u>, for two reasons: (1)
the proposed values were derived from the same body of current data and
current theories which are to some extent (perhaps to a considerable
extent) uncertain; (2) it is to be expected that authors making later
estimates were aware of earlier estimates and were influenced by them,
so that the estimates given in Table 1 are not really independent.

Table 1

	a	b	c	d	e
ρ	1	1.1	1.3	1.0	1.1 ± 0.1
ϕ_p	0	0	-0.3	-0.3	-0.1 ± 0.2
$\tilde{\omega}_e$	0	-0.5	0	0.5	0 ± 0.3
ϕ_ℓ	0	0	-0.7	0	-0.2 ± 0.3
ϕ_i	-1	0	0	0	-0.2 ± 0.4
ϕ_c	-1	-0.3	-0.3	-2	-0.9 ± 0.7
λ	2-8	6	8	7	6.3 ± 1.9

Estimates of quantities occurring in the text.
Estimates <u>a,b,c</u>, and <u>d</u> are taken from Shklovskii
and Sagan (1966), Cameron (1963), Billingham and
Oliver (1973), and Sagan (1974), respectively. Values
in column <u>e</u> are derived from estimates <u>a</u> to <u>d</u>.

Using the estimates of $\bar{\rho}$, $\sigma(\rho)$, etc., given in the last column of
Table 1, we may estimate $\bar{\nu}$ and $\sigma(\nu)$ by using equations (6) and (7). We
find that $\bar{\nu} \approx 6$ and $\sigma(\nu) \approx 2$. We see that the 1σ range of values of N
is 10^4 to 10^8. That is, on the basis of the information presented, we
have only 70% confidence that N lies in the range 10^4 to 10^8. We may
have 95% confidence that N lies in the 2σ range which is seen to be 10^2
to 10^{10}. On remembering that our estimate of the range is <u>conservative</u>,
we see that there is an enormous uncertainty in current estimates of N.

We see from Table 1 that 80% of the variance of ν, $\sigma^2(\nu)$, is due to
our uncertainty of the "lifetime" λ, and more than half of the remainder
of the variance is due to the uncertainty of ϕ_c. This is not surprising,
since these are the most speculative estimates involved in the Drake
formula. As Bracewell (1978) has pointed out, it is a gross simplifi-
cation to think of a single mean lifetime for civilizations. Bracewell
recommends that we consider the division of civilizations into groups.
It certainly makes sense to try to expose the major assumptions underlying estimates of L and also to inquire into how these assumptions
influence the probability that a civilization will establish radio
beacons or be a source of "leakage" radio emission, since these are the

two possibilities which must be considered in assessing whether or not
it makes sense to conduct a search for extraterrestrial intelligent
radio signals. This question will be discussed in the next section.

2. ESTIMATES OF THE COMMUNICATIVE LIFETIME

Much of present-day discussion of the possible existence of advanced
civilizations in the Galaxy hinges on the possibility that such civilizations might be detected through their radio emission. (See, for
instance, Morrison, Billingham and Wolfe, 1977.) From a certain viewpoint, this makes good sense. For instance, if scientists were to be
given the definite charge of searching for extraterrestrial civilizations, they would have no choice but to carry out their search using
known physical principles and current or "accessible" technology. There
is no doubt that, if we were to try to send signals from earth to civilizations many light-years away, we would use radio transmission for
reasons which have been thoroughly explored and persuasively presented.

However, if scientists are instead attempting to assess the likelihood that a contemplated search may be successful, or if they are trying
to compare the prospects of success of two or more strategies, then we
must face the possibility that civilizations much more advanced than
our own (more advanced perhaps by many millions of years) may use
communication technologies far superior to those we know, based on
physical principles of which we are now utterly ignorant. This concept
will be described in shorthand form as the proposition that there exists
a "hyperphysics" of which we are now ignorant. As one possibility, this
hypothesis would include the case that our familiar four-dimensional
space-time is really a section of a hyperspace, and that it is possible
to obtain access to other sections of this hyperspace by technological
means. Since our known laws of physics refer only to the familiar
four-dimensional space, we have no reason to believe that familiar
limitations of travel time, etc., would have any relevance to such a
hyperspace. Clearly, if an advanced civilization discovers a way to
send messages at speeds much greater than the speed of light, radio
waves would not be used for interstellar communication.

However, consideration of a possible "hyperphysics" carries with
it even more profound implications for the SETI debate. According to
present-day physics, interstellar travel would be very slow and extraordinarily expensive in energy and money. (See, for instance, Marx,
1973.) In that same volume, Kardashev (1973) writes "In dealing with
extraterrestrial intelligence, we must concern ourselves with certain
definite models; if we are considering a model of a super civilization,
that is, a civilization that is far ahead of ours, in looking for it we
must take into account things we know nothing about. Many people think
that nowadays in astrophysics we know a great deal about all objects.
In my opinion this is not so at all." Kardashev goes on to consider
the possibility that it will at some time be possible to pass from one
"space" to another, basing his discussion on present-day theories of
black holes. These opening remarks by Kardashev were followed by exten-

sive discussion involving Ginzburg, Gold, Sagan, Townes and von Hoerner about the possibility that there exist as yet undiscovered laws of physics. The possibility that we live in a hyperspace, or that there exists some other form of still undiscovered hyperphysics, has profound implications for discussion of interstellar travel as well as for discussion of interstellar communication.

In recent years, several authors (Bracewell, 1974; Schwartzman, 1977; Kuiper and Morris, 1977; Jones, 1976, 1978) have considered the concept of "Galactic Colonization." Even using means of space travel which are consistent with present-day physics, it has been argued that if a single civilization were to develop even to our current level of technological sophistication, it would be only a matter of time (perhaps one million years) before all habitable planets in the Galaxy would be colonized. This argument is taken sufficiently seriously that the absence of obvious evidence that we are a colony is taken to imply that we are the only advanced civilization in the Galaxy (Hart, 1975). If colonization is possible or likely even with present-day physics, how much more likely it must be for any civilization which discovers a hyperphysics and the means to exploit it.

It appears from the preceding discussion that the fundamental question underlying consideration of a search for extraterrestrial civilizations or communication with them is whether or not it is possible that advanced civilizations will discover a hyperphysics such as the discovery that we live in a hyperspace and the discovery of techniques to navigate that hyperspace. We therefore introduce the following symbol:

H = actuality of hyperphysics
$\overline{H}$ = no actuality of hyperphysics (limitations of current physics are absolute)

Since, for reasons which will become apparent, it will be necessary to use a more complex notation for probabilities, we shall in this section not work in terms of probability distributions although it is implicitly recognized that such distributions are necessary. As we saw in Section 1, we may give a simple representation of the uncertainty of any estimate of a probability by ascribing a standard deviation to the estimate of the probability or, better, to the estimate of the logarithm of the probability or of the "odds" defined by $P/(1-P)$ (Good, 1950).

We shall denote by $(A|B)$ the probability that proposition A is true on the basis of proposition B (Sturrock, 1973). Since our assessment of the probability of H is based more on ignorance than on knowledge, it will be denoted by $(H|-)$ and the probability of $\overline{H}$ by $(\overline{H}|-)$.

The next consideration with important implications for a search strategy appears to be whether or not a Galactic Federation (or, as Bracewell [1974] calls it, a "Galactic Club") comes into existence. We will denote by G the proposition that a Galactic Federation exists and by $\overline{G}$ the proposition that it does not. Since rapidity of communication and travel would promote exploration and peaceful or nonpeaceful visitation, it seems that $(G|H)$, the probability of there being a Galactic Federation in the hyperphysics scenario, would be much larger than $(G|\overline{H})$, the probability of there being a Galactic Federation if the limitations of present-day physics are absolute.

We can now distinguish four expected lifetimes for advanced civilizations: $L(G,H)$ (the expected lifetime [in years] if H is true and if G is true), $L(G,\overline{H})$, $L(\overline{G},H)$ and $L(\overline{G},\overline{H})$. By including these estimates in equation (1), we arrive at $N(G,H)$, the expected number of advanced civilizations in the Galaxy if both G and H are true, $N(G,\overline{H})$, etc. It will be convenient to write

$$N(G,H) = K\ L(G,H), \text{ etc.,} \tag{8}$$

where

$$K = R f_p\ P_e f_\ell f_i f_c. \tag{9}$$

We denote by $(R_B|G,H)$ the probability that, if both G and H are true, an advanced civilization will operate radio beacons; similarly for $(R_B|G,\overline{H})$, etc. Then $N(R_B)$, the expected number of civilizations operating radio beacons in the Galaxy, is given by

$$N(R_B) = K\left[(R_B|G,H)\tilde{L}(G,H) + (R_B|\overline{G},H)\tilde{L}(\overline{G},H)\right. \tag{10}$$
$$\left. + (R_B|G,\overline{H})\tilde{L}(G,\overline{H}) + (R_B|\overline{G},\overline{H})\tilde{L}(\overline{G},\overline{H})\right]$$

in which

$$\tilde{L}(G,H) = (G,H|-)\ L(G,H), \text{ etc.} \tag{11}$$

where

$$(G,H|-) = (G|H)\ (H|-), \text{ etc.} \tag{12}$$

Similarly, if R_L denotes the leakage of "domestic" radio waves, the expected number of advanced civilizations leaking radio waves is given by

$$N(R_L) = K\left[(R_L|G,H)\tilde{L}(G,H) + ...\right]. \tag{13}$$

It has been argued persuasively on many occasions (see, for instance, Wolfe, 1977) that on the basis of present-day physics ($\overline{H}$) "manned interstellar flight is out of the question not only for the present but for an indefinitely long time in the future." For these reasons, most discussions of search strategies for extraterrestrial life, being based on ($\overline{H}$), ignore the possibility that advanced civilizations may undertake exploration and, subsequently, "manned" surveillance (S_M) by means of space vehicles. However, Bracewell (1974) has pointed out that, even within the context of $\overline{H}$, it is possible for an advanced civilization to carry out surveillance by artificial means such as surveillance by means

of a space vehicle "parked" near a star such as the sun, and equipped
to detect radio signals (S_R). Bracewell conceives that, once the probe
has detected radio signals, it immediately engages in open dialogue
with terrestrial civilizations. However, this scenario runs counter
to the practice of all terrestrial intelligence organizations, which
set out to learn as much as possible about other societies but divulge
no real information although they may disseminate disinformation. Since
a radio probe could learn a great deal simply by listening and runs the
risk of being captured if it transmits radio signals from near the earth,
it seems much more likely that, for some considerable time, such a probe
would merely maintain radio surveillance, transmitting whatever it learns
back to its home base.

Formulas analogous to equations (10) and (13) give estimates of
the number of advanced civilizations in the Galaxy practicing surveillance either by "manned" vehicles or by radio means:

$$N(S_M) = K\left[(S_M|G,H)\tilde{L}(G,H) +\ldots\right] \tag{14}$$

$$N(S_R) = K\left[(S_R|G,H)\tilde{L}(G,H) +\ldots\right] \tag{15}$$

One could imagine that such probes may search for signals of a
type we are not now using, but this consideration would not figure in
present-day strategy for extraterrestrial intelligent life and so may
be ignored.

The above formalism can be helpful in laying out the possible
scenarios which might lead to detectable radio signals and for seeing
what judgments are involved in estimating probabilities of these
scenarios. We also see that, even though many factors are involved in
estimating each quantity $N(R_B)$, etc., the quantities which have been
combined together as K (equation 9) will cancel out in comparing one
number with another, e.g., $N(R_B)$ with $N(R_L)$.

In order to make numerical estimates of these quantities, it is
desirable that a number of scientists should make estimates of the
various quantities involved, so that one could then assign a value with
an error bar to each quantity, as was done in Table 1 of Section 1. It
is hoped that this can be done at a later date. However, at this time,
the best that I can do is to present my own estimates of the various
quantities and the resulting estimates of $N(R_B)$, etc.

The quantity K may be derived from Table 1. It is found that

$$\log K = 10^{-0.3\pm1}. \tag{16}$$

The most difficult estimate to make is that of $(H|-)$ and $(\overline{H}|-)$. On the
one hand we clearly have no reason to believe that there is a "hyperphysics" with laws transcending those in current use. On the other
hand, science as we know it dates back only about 2,000 years; the laws

of gravitation and motion have been known for only about 300 years, electromagnetism for about 100 years, and quantum theory and relativity for only about 50 years. Why should we believe that, if scientists were to continue working for another million years, there would not be comparable revolutions or revelations? My own attitude is to assign maximum uncertainty to the proposition H by assigning to its probability a uniform distribution over the range zero to unity. With this distribution, we find that

$$(H|-) = (\overline{H}|-) = 10^{-0.4\pm0.4}. \tag{17}$$

We must next consider the probability of the existence of a "Galactic Federation." Within the context of hyperphysics, which is assumed to facilitate rapid communication and travel, it seems more likely than not that there would be a Galactic Federation. However, this issue is so speculative that the following very conservative estimates will be adopted:

$$\left[(G|H)/(\overline{G}|H)\right] = 10^{1\pm1} \tag{18}$$

If, on the other hand, there is no hyperphysics so that communication is limited by the speed of light and travel is very slow and extraordinarily expensive, the development of a federation with enforceable conformity seems highly unlikely. We therefore adopt

$$\left[(G|\overline{H})/(\overline{G}|\overline{H})\right] = 10^{-4\pm2}. \tag{19}$$

It follows from these equations that, to accuracy sufficient for our purposes, $(G|H) \approx 1$ and $(\overline{G}|\overline{H}) \approx 1$.

We next consider the probable lifetime L for each of the four cases. Case $(\overline{G},\overline{H})$ is the one implicitly considered in Section 1, so that we may use the estimate given in Table 1, $L(\overline{G},\overline{H}) = 10^{6\pm2}$. Within the context of $\overline{H}$, it is unlikely that the existence of a Galactic Federation would make much difference, so that we may adopt also $L(G,\overline{H}) = 10^{6\pm2}$. Within the context of H, the existence of a Galactic Federation may be very effective in preserving civilizations, so that we may adopt the more optimistic estimate $L(G,H) = 10^{8\pm2}$. Within the context of H, but in the absence of a Galactic Federation, civilizations may be as precarious as they are in the context of $\overline{H}$, so that we may again set $L(\overline{G},H) = 10^{6\pm2}$.

The probabilities $(G,H|-)$ etc., given by equation (12), are listed in column 1 of Table 2. The lifetimes L are listed in column 2, and the "reduced" lifetimes, given by equation (11), are listed in column 3. From these estimates, (G,H) appears to be the most important possibility, followed by $(\overline{G},\overline{H})$, $(\overline{G},H)$ and $(G,\overline{H})$. This puts us in the uncomfortable situation that the most important scenario appears to be that of which we know least.

We must now turn to estimates of the probatilities $(R_B|G,H)$, etc.

If the limitations of present-day physics are absolute ($\overline{H}$), civilizations
are not subject to easy attack from their interstellar neighbors, yet
there is some interest and possible advantage in communication. It seems
therefore that the establishment of radio beacons would not be unlikely.
Nor would there be any harm in allowing for radio leakage. On the other
hand, it would still be prudent to begin by radio surveillance. "Manned"
surveillance would be exceedingly difficult. If there is a Galactic
Federation, there may be somewhat more interest in setting up beacons
and in arranging for radio surveillance, but these tasks may be assigned
to a fraction of the civilized worlds. In any case, the existence of a
Galactic Federation would not do much to facilitate manned surveillance.
For these reasons, estimates for the probability of R_B, etc., on the
assumption $\overline{H}$, are taken to be independent of the alternatives $G, \overline{G}$. The
values are listed in columns 4 through 7 of Table 2.

Table 2

		P	L	$\tilde{L}$	R_B	R_L	S_M	S_R
G	H	-0.4 ± 0.4	8 ± 2	7.6 ± 2	-4 ± 2	-2 ± 2	-1 ± 1	-4 ± 2
$\overline{G}$	H	-1.4 ± 1	6 ± 2	4.6 ± 2	-4 ± 2	-2 ± 2	-1 ± 1	-4 ± 2
G	$\overline{H}$	-4.4 ± 2	6 ± 2	1.6 ± 3	-1 ± 1	-1 ± 1	-5 ± 2	-0.5 ± 0.5
$\overline{G}$	$\overline{H}$	-0.4 ± 0.4	6 ± 2	5.6 ± 2	-1 ± 1	-1 ± 1	-5 ± 2	-0.5 ± 0.5
N					4.5 ± 2.5	5.5 ± 3	6.5 ± 2.5	4 ± 2

In the column headed P are listed the logarithms of $(G, H|-)$, etc.;
under L, the logarithms of the lifetimes $L(G, H)$, etc.; under $\tilde{L}$,
the logarithms of $\tilde{L}(G, H)$, etc; and under R_B, etc., the logarithms
of the probabilities $(R_B|G, H)$, etc. Also under R_B, etc., are the
final estimates of $N(R_B)$, etc.

    If, on the other hand, there exists a hyperphysics which we have
yet to discover but which is available for exploitation by more advanced
civilizations, making possible rapid communication and interstellar
travel, the strategies are likely to be quite different leading to quite
different estimates for the probabilities of R_B, etc. If manned surveillance is possible, there would be little reason to set up radio
beacons or to establish probes equipped for radio surveillance. Hence
the probability of S_M is likely to be high and the probabilities of R_B
and S_R are likely to be low. It is quite possible that electromagnetic
waves will be superseded as a mechanism for domestic communication, in
which case there would be no leakage. However, electromagnetic waves
may be retained for certain specialized purposes: the question then is
whether the civilization would or would not take pains to suppress this
evidence of its existence. If there is no Galactic Federation, each
civilization may be somewhat wary of advertising its existence and
location. If there is a Galactic Federation, each civilization may
feel sufficiently secure to allow for radio leakage. In scenario (G, H),

radio leakage would seem somewhat more likely than radio beacons or
radio probes. Since it is not clear how the existence of the Galactic
Federation would influence the probabilities of R_B, etc., these are
taken to be the same for the two cases G,$\bar{G}$. The proposed values are
listed in Table 2.

On using equations (10), etc., and the values given in the first four
rows of Table 2, we may estimate $N(R_B)$, etc. These values are given in
row 5 of Table 2.

3. DISCUSSION

It is not the purpose of this article to present and defend any
particular set of numerical estimates of $N(R_B)$, etc. The aim is, rather,
to present a formalism which can be used as a "bookkeeping procedure"
allowing us to list what appear to be possible scenarios leading to
possible communication with extraterrestrial civilizations and
to identify the decisions which must be made in estimating the probability of any particular mode of contact, and which shows how the various
decisions may be combined to arrive at the final estimates. A further
goal is to show how one may represent the uncertainty in each decision
and carry it through to the final estimates.

I have argued that the most important decision to be made is
whether one considers it likely that the laws of present-day physics
are absolute or whether one considers that there exists an as-yet-
undiscovered "hyperphysics," possibly involving hyperspace, which makes
it possible to escape from the limitations of physics as it is now
understood.

It may well be argued that this is not a scientific question, since
science deals with what we know about the universe, to which one might
respond that in any scientific discipline an assessment of our ignorance
is just as important as an assessment of our knowledge. However, it
seems to me that the question does lie outside science; if it belongs
to any discipline at all, it is likely to be the philosophy of science.
Since it is unlikely that the exponents of this branch of learning have
any way of arriving at a concensus on this question, one must leave the
probability H as a purely subjective estimate. Fine (1973), in discussing the nature of probability statements in discussions of the preva-
lence of extraterrestrial intelligent life, presents his judgment that
"the concept of subjective probability is at present the only basis upon
which probability statements can be made about ETIL [extraterrestrial
intelligent life]."

Even though we cannot expect conformity in estimates of the various
probabilities figuring into our final estimates, it is possible that we
can solicit estimates from a number of scientists and then represent
their collective judgment by means of a distribution function, as indicated in Section 1. In this way we will at least be able to represent

the "collective subjective estimates" of interested scientists, rather
than just one scientist's conjectures.

Given the preceding caveats, we may now turn to the estimates presented in Table 2. In comparing the probable numbers of civilizations
operating radio beacons or allowing radio leakage, we see that the
latter appears to be larger, but not by a statistically significant
amount. What is more significant is that the principal contribution to
$N(R_B)$ comes from the case $(\overline{G},\overline{H})$ and the principal contribution to $N(R_L)$
comes from the case (G,H).

The number of civilizations likely to be operating radio probes
seems to be almost as large as those operating beacons or allowing leakage, substantiating Bracewell's (1960) proposal that we search for such
devices. The principal contribution to our assessment of $N(S_R)$ comes
from the case $(\overline{G},\overline{H})$.

We see, however, that the assumptions set out in the first four
lines of Table 2 lead to an estimate of $N(S_M)$ somewhat larger than the
others. The question which now arises is the following: if we are
under active surveillance by manned craft, would we know it? Hart
(1975) asserts categorically that we now are not being visited. Most
scientists involved in the SETI program appear to agree with him. However surveillance, even within our terrestrial civilization, can be
covert and hard to detect--due not merely to paucity of public information but also to the dissemination of disinformation. This implies
that the assessment of whether or not we are under covert surveillance
by extraterrestrial civilizations is not a purely scientific question--
it spills over into scientific intelligence.

Although we cannot infer from present-day physics and astronomy
that we are--or are not--under covert surveillance by extraterrestrial
civilizations, we can place some credence in the following argument. If
we are under manned surveillance, the transportation is not likely to
be effected by spacecraft as we know them (Markowitz, 1967). The fact
of manned surveillance (if it is ever established) would in itself argue
for the existence of "hyperspacecraft" which would in turn imply the
actuality of a hyperphysics (Sturrock, 1975, 1978).

It is a pleasure to acknowledge stimulating comments on
an early draft of this article from R.N. Bracewell, F.D. Drake,
P. Morrison, C. Seeger and S. Von Hoerner. This research was
supported in part by the National Aeronautics and Space Administration under grant NGR 05-020-668.

REFERENCES

Billingham, J., and Oliver, B.M.: 1973, Project Cyclops (NASA Ames
     Research Center Report CR 114445), p. 26.

Bracewell, R.N.: 1960, Nature, 186, p. 670.

Bracewell, R.N.: 1974, The Galactic Club: Intelligent Life in Outer
    Space (Stanford Alumni Association, Stanford, Calif.).

Bracewell, R.N.: 1978, Acta Aeronautica, 6, pp.67-69.

Cameron, A.G.W.: 1963, Interstellar Communication (A.G.W. Cameron, ed.,
    Benjamin, New York), pp.309-315.

Fine, T.: 1973, Communication with Extraterrestrial Intelligence
    (C. Sagan, ed., MIT Press, Cambridge, Mass.), pp.357-361.

Good, I.J.: 1950, Probability and the Weighing of Evidence (Griffin, London)

Hart, M.H.: 1975, Q.J.R.A.S., 16, pp.128-135.

Jones, E.M.: 1976, Icarus, 28, pp.421-422.

Jones, E.M.: 1978, J. Brit. Interplan. Soc., 31, pp.103-107.

Kardashev, N.S.: 1973, Communication with Extraterrestrial Intelligence
    (C. Sagan, ed., MIT Press, Cambridge, Mass.), p.192.

Kuiper, T.B.H., and Morris, M.: 1977, Science, 196, pp.616-621.

Markowitz, W.: 1967, Science, 157, pp.1274-1279.

Marx, G.: 1973, Communication with Extraterrestrial Intelligence
    (C. Sagan, ed., MIT Press, Cambridge, Mass.), p.216.

Morrison, P., Billingham, J., and Wolfe, J.: 1977, The Search for
    Extraterrestrial Intelligence (SP-419, NASA, Washington, D.C.).

Sagan, C.: 1974, Interstellar Communication: Scientific Perspectives
    (C. Ponnamperuma and A.G.W. Cameron, eds., Houghton Mifflin,
    Boston), pp.1-24.

Schwartzman, W.T.: 1977, Icarus, 32, pp.473-475.

Shklovskii, I.S., and Sagan, C.: 1966, Intelligent Life in the Universe
    (Holden and Day, San Francisco), Ch.29.

Sturrock, P.A.: 1973, Astrophys. J., 182, pp.569-580.

Sturrock, P.A.: 1975, Stanford Workshop on Extraterrestrial Civiliza-
    tions: Opening a New Scientific Dialog (J.B. Carlson and P.A.
    Sturrock, eds., Origins of Life, 6, pp.459-470), p.469.

Sturrock, P.A.: 1978, Q.J.R.A.S., 19, pp.521-523.

Wall, J.V.: 1979, Q.J.R.A.S., 20, pp.138-152.

Wolfe, J.H.: 1977, The Search for Extraterrestrial Intelligence
    (P. Morrison, J. Billingham, and J. Wolfe, eds., SP-419, NASA,
    Washington, D.C.), p.107.

A NEW APPROACH TO THE NUMBER N OF ADVANCED CIVILIZATIONS IN THE GALAXY*

V.S. Troitskii
Radiophysical Research Institute
Gorky, U.S.S.R.

ABSTRACT   The nearly simultaneous origin of life everywhere in our galaxy, and possibly in the entire Universe, is proposed as an alternative to the generally accepted concept of a continuous appearance of life. The impulsive appearance of life provides a better explanation to the possible absence of advanced civilizations in our galaxy.

1. THE ESTIMATION OF N

Practically all efforts to estimate the number of technologically advanced civilizations in our galaxy have been based on the famous Drake equation (Drake, 1965; Shklofskii and Sagan, 1966) which assumes that life in the Universe arises continuously following the formation of planets with suitable conditions. An alternative approach proposed here is to assume that life originated only once in the galaxy, and possibly in the whole Universe, and that this event occurred in a narrow time interval that corresponds to only a small fraction of the age of the galaxy. It is also assumed that life arose only on those planets where at the time of this event the conditions necessary for life had already developed. Life did not arise anywhere else, neither did it arise at any earlier or at any later time than this impulsive event.

This new approach might seem somewhat arbitrary, but since it does not contradict any physical laws and since there is no evidence in support of the continuous appearance of life, it should also be considered as a viable alternative. Actually there is no reason why we should not consider an impulsive origin of life when we have accepted the Big Bang origin of the Universe and we have essentially abandoned the competing theory of a continuous creation of matter.

---

* This paper was abstracted by the Editor from a lengthier paper by Dr. Troitskii which arrived in a draft form as this Volume was going to press.

M. D. Papagiannis (ed.), Strategies for the Search for Life in the Universe, 73–76.
Copyright © 1980 by D. Reidel Publishing Company.

The simultaneous origin of life might be related to the evolution of the Universe as a whole, or of our galaxy in particular such as the timing of the formation of earth-like planets around sun-like stars following a long period of nucleosynthesis of the heavy elements in the galaxy. A different explanation, which leads to essentially the same result, has been discussed by P.V. Makovetskii. In this theory, life had originated and had already evolved to advanced forms in many planets of the galaxy prior to this event. All advanced forms of life, however, perished 4-5 billion years ago following a cataclysmic catastrophe, possibly the burst from the galactic center, that affected the entire galaxy. Only the simplest forms of life managed to survive this holocaust and from that time on, life began again to evolve in the different solar systems of the galaxy with an essentially simultaneous start.

2.   THE ABUNDANCE OF CIVILIZATIONS IN THE TWO APPROACHES

At the time of the simultaneous origin of life, the number N_p of suitable planets in the galaxy where life could originate was,

$$N_p = R_p \tau_p \tag{1}$$

where R_p is the rate at which planets suitable for life are formed in the galaxy, and τ_p is the time period over which they manage to retain these favorable to life conditions, such as liquid water on their surface. It should be pointed out that N_p is likely to be considerably smaller than the total number of planets suitable for life that were formed throughout the history of the galaxy. Consequently the number N_ℓ of planets where life originated in the simultaneous origin is likely to be much smaller than the number of planets where life originated in the continuous origin.

Following the impulsive origin of life, which from the known case of the earth must have occurred roughly 4 billion years ago, life began on each planet its independent evolution, which in a number of cases has led, or is bound to lead to advanced communicative civilizations. The evolutionary period, τ_ℓ, for the appearance of advanced civilizations will most likely vary from case to case (Kreifeldt, 1971), forming probably a Gaussian distribution around a mean value $\overline{\tau}_\ell$ with a standard deviation $\alpha \overline{\tau}_\ell$, where $\alpha < 1$. Let τ_0 be the time that has elapsed since the simultaneous origin of life, which must be very close also to the evolutionary period to an advanced civilization in the case of the earth. The number N of advanced civilizations currently present in the galaxy will depend on two factors: the relation of τ_0 to $\overline{\tau}_\ell$, and the life expectancy L of advanced civilizations.

If $\tau_0 \ll \overline{\tau}_\ell$, very few civilizations will have reached by now the communicative stage, while if $\tau_0 > \tau_\ell$ at least half of the maximum

possible number of cases will have already reached this stage. The
effect of L is in general to reduce the number of civilizations currently present as compared to the number that have appeared throughout
the past history of the galaxy. If $L = \infty$, the two numbers are the
same because all the old civilizations are still active. If $L < \tau_0$,
however, the number present is smaller than the total because some of
the older civilizations have long disappeared.

The differences between the two approaches are the following:
For $L = \infty$, N in the simultaneous origin will in the long run be much
smaller than in the continuous origin, because, as discussed above,
the number of planets that can host life in the simultaneous origin is
much smaller. In addition, in the continuous origin many more older
civilizations will be present, simply because in this approach life
began to form and evolve much earlier. For a limited value of L, in
the continuous origin N after an initial rise tends to stabilize at a
constant value, while in the simultaneous origin N reaches a peak and
then decreases back to zero. In addition, in the continuous origin
the number of civilizations that have ceased to exist is likely to be
100-1000 times larger than N (Freeman and Lampton, 1975), a result
with important consequences as we will see in the next section.

Another important difference between the two approaches is their
sensitivity to differences in the evolutionary periods τ_0 and $\overline{\tau_\ell}$. If
τ_0 is considerably smaller than $\overline{\tau_\ell}$, in the continuous origin there
will still be a substantial number of civilizations present, because
the process has been going on for a much longer time and life had
ample time to evolve into advanced civilizations in many more planets.
In the simultaneous origin, however, because of the shortness of time
there will be very few advanced civilizations present and practically
all of them will be very young.

3.   COMPARISONS WITH EXPERIMENTAL EVIDENCE

Almost four decades of radioastronomical observations and two decades of about a dozen special searches in the USSR, USA and Canada
for radio signals from extraterrestrial intelligence, have not produced any evidence that there are any other communicative civilizations in our galaxy. Of course these searches have covered only a
very small fraction of the search space available, both in terms of
stars and in terms of frequencies, and therefore we may very easily
have missed them. In addition, the absence of powerful omnidirectional
radio beacons operated by advanced galactic civilizations can be explained by difficulties these beacons will impose on their energy and
material resources and by their need to preserve their interplanetary
medium (Troitskii, 1980).

Still it begins to look that our galaxy is by no means full of
advanced civilizations, and as a matter of fact, as suggested also by
I.S. Shklovskii, we might be all alone in the galaxy. A small value

for N can be explained by the Drake equation by postulating a low value for L, but this explanation requires also that a large number of advanced civilizations had been present during the long history of our galaxy. This, however, is not supported by present evidence because, as it was pointed out by Hart (1975), some of these old civilizations would have initiated interstellar travelling and the entire galaxy would have been colonized a long time ago. There is no evidence, at least on earth, that the colonization of the galaxy has already taken place, though further exploration of our solar system and in particular of the asteroid belt (Papagiannis, 1978) is needed to make this argument stronger.

All in all, the presently available evidence points toward a very low value of N and to the absence of large numbers of advanced civilizations in the past history of our galaxy. These results are not compatible with the commonly presented conclusions of the Drake equation (Goldsmith and Owen, 1980), though they could still be explained by the continuous origin if, e.g., planets capable of maintaining the conditions needed by life for periods longer than $\overline{\tau}_\ell$ were extremely rare in the galaxy. The evidence discussed above, on the other hand, is totally compatible with a simultaneous origin in which τ_0 is considerably smaller than $\overline{\tau}_\ell$. Under these conditions our planet would be one of the very first in the galaxy where life has evolved into an advanced technological civilization and hardly any other civilizations would have appeared in the past history of our galaxy. In conclusion we see that the impulsive origin of life approach appears to be in very good agreement with the presently available evidence.

ACKNOWLEDGEMENTS

I would like to thank P.V. Makovotskii and L.M. Ginzburg for their helpful discussions and their valuable comments and suggestions.

REFERENCES

Drake, F.D.: 1965, Current Aspects of Exobiology, ed. Mamikunian, G. and Briggs, M.H., Pergamon Press.
Freeman, J. and Lampton, M.: 1975, Icarus, 25, 368.
Goldsmith, D. and Owen, T.: 1980, The Search for Life in the Universe, The Benjamin/Cummings Publ. Co.
Hart, M.H.: 1975, Q.Jl.R. Astr. Soc., 16, 128.
Kreifeldt, J.C.: 1971, Icarus, 14, 419.
Papagiannis, M.D.: 1978, Q.Jl.R. Astr. Soc., 19, 277.
Shklovskii, I.S. and Sagan C.: 1966, Intelligent Life in the Universe, Holden Day, San Francisco.
Troitskii, V.S.: 1980, Zemlya i Vselennaya.

PART II

STRATEGIES FOR SETI THROUGH RADIO WAVES

STRATEGIES FOR SETI THROUGH RADIO WAVES
AN INTRODUCTION

Bernard M. Oliver
Hewlett-Packard Company

A peculiar opposition has developed to SETI. It is argued that, if intelligent civilizations were as common as SETI enthusiasts believe, Earth surely would have been colonized or at least visited long ago. Of course, followers of Von Däniken believe we <u>have</u> been, while the UFO buffs believe we are <u>still</u> being visited. But a more skeptical faction, finding no solid evidence either for visitation or for astro-engineering on a Dyson scale, concludes that either we are being quarantined (the zoo hypothesis) or else intelligent life is a very rare phenomenon (we really <u>are</u> alone).

It seems to me that both of these conclusions are entirely unwarranted; they attach far too much credence to that fantastic extrapolation, the Kardashev type II civilization, to interstellar travel generally, and to interstellar colonization in particular. The energies required to construct Dyson spheres are, in a word, astronomical. To disassemble an Earth-sized planet requires as much energy as in all the sunlight that has fallen on Earth since the days of the dinosaurs, 70 million years ago. To produce four velocity increments of $0.7c$ in an ideal rocket requires the <u>annihilation</u> of 16 times the mass of the payload, or enough energy to power the United States with electricity for over 100,000 years. Having spent several years trying to get approval of even a modest radio SETI, I have great difficulty visualizing <u>their</u> congress approving a single mission costing this much energy (let alone a flotilla to explore the Galaxy) or, indeed, any slower, less energetic flights, where the results would be unknown to the launching generation. Years of science fiction have brainwashed us. It seems almost certain to me that there is no extensive interstellar travel.

This leaves radio as the only rational alternative for SETI. Today we will hear about our first feeble attempts at radio SETI both in the U.S. and the USSR, and about some future plans. I use the adjective feeble not in a pejorative sense, but rather because our listening devices may have to be made much more sensitive before we succeed. However, since we do not <u>know</u> this to be true and, since great sensitivity is expensive, our first attempts should be feeble.

M. D. Papagiannis (ed.), Strategies for the Search for Life in the Universe, 79–80.
Copyright © 1980 by D. Reidel Publishing Company.

To me one of the most impressive aspects of radio SETI today is that significant programs may be about to begin with government support both in the USSR and in the U.S. That implies a spirit of scientific inquiry about the universe, indeed a state of intellectual maturity in our society, that is quite surprising for a world in which wars still exist.

MICROWAVE SEARCHES IN THE U.S.A. AND CANADA

B. Zuckerman
University of Maryland

Jill Tarter
University of California, Berkeley

Search strategies for ETI may run the gamut from passive, in which they do all the work, to active, in which we do all the work. Searches for electromagnetic radiation occupy a middle ground – success requires a cooperative effort. Indeed the Cyclops[1] design was based in part on the idea of cost-sharing between the transmitting and receiving societies.

To date, there have been fewer than 20 radio searches for signals from ETI. All have, apparently, produced negative results unless someone is hiding something in their bottom drawer. We will discuss six of the most sensitive searches that have been carried out in the United States of America and in Canada.

All six of the searches listed in Table 1 assume that interstellar communication will involve a signal of very narrow bandwidth. If the transmitting society has only a certain amount of microwave power available to transmit, then the signal-to-noise ratio at the receiving end can be made highest when this power is concentrated into the narrowest possible range of frequencies. In addition, very narrow signals can easily be distinguished from ordinary thermal emission from interstellar molecules and atoms. Signposts for signals from ETI are, therefore, either rapid variability (timescales of seconds to days) or a bandwidth significantly narrower than either a thermal bandwidth corresponding to temperatures ~ 10 K, or a line narrowed unsaturated maser.

Patrick Palmer and Ben Zuckerman (hereafter PZ) were the first to take advantage of the very considerable advances in instrumentation in spectral line radio astronomy in the 1960's. They pointed the 91-meter transit telescope of the U.S. National Radio Astronomy Observatory toward stars within 25 parsecs (80 light years) of the Sun. These stars were taken mostly from Gliese's catalog[2] plus a few from the RGO catalog[3]. The motivation for a targeted search of nearby stars (four of the six programs in Table 1) is as follows. Given existing telescopes and receivers, we are not yet capable of detecting omni

81

M. D. Papagiannis (ed.), Strategies for the Search for Life in the Universe, 81–92.
Copyright © 1980 by D. Reidel Publishing Company.

TABLE 1

RADIO SEARCH PROGRAMS

DATE	OBSERVER	OBSERVATORY	WAVELENGTH	TARGET	SENSITIVITY (W/SQ M)
1972 – 76	PALMER ZUCKERMAN	NRAO 91 m	21 cm	~ 670 NEARBY STARS	10^{-23}
1973 – PRESENT	DIXON COLE	OHIO STATE 53 m	21 cm	ALL SKY	FEW $\times 10^{-21}$
1974 – LIMBO	BRIDLE FELDMAN	ALGONQUIN 46 m	1.3 cm	~ 70 NEARBY STARS	FEW $\times 10^{-22}$
1975 – 76	DRAKE SAGAN	ARECIBO 305 m	21 cm 18 cm 12.5 cm	SEVERAL NEARBY GALAXIES	10^{-24}
1977	TARTER AND FRIENDS	NRAO 91 m	18 cm	200 NEARBY STARS	FEW $\times 10^{-24}$
1978	HOROWITZ	ARECIBO 305 m	21 cm	185 NEARBY STARS	FEW $\times 10^{-27}$

directional broadcasts from Kardashev Type I civilizations across
interstellar distances (a Type I civilization is one that is not much in
advance of our own). Therefore, if we are to detect Type I civilizations,
we must look for powerful beamed transmissions. Such beacons are not
likely to be pointed at Earth except by our nearby neighbors.

Targeted searches that examine only a few hundred nearby stars are
not particularly significant in the best-guess Drake/Sagan[4] scenario –
$\lesssim 10^6$ civilizations in the Milky Way – since $\gtrsim 10^5$ stars should be
searched to find one civilization. But if the Milky Way has been
physically "colonized" by means of interstellar rocket travel, then
many (most) of the nearby star systems should harbor technical civilizations and the negative results to date may already be of some
significance[5].

PZ limited their search to F, G, K, and M type main sequence stars
based on classical arguments: such stars may provide a habitable zone
which is stable for a sufficiently long time to allow life to originate
and to evolve into a technological civilization. In the absence of
interstellar colonization, these limits on spectral type may be much too
broad if the continuously habitable zone is as narrow as estimated by
Hart[6]. But, if the galaxy has already been colonized, then essentially
all main sequence stars, and possibly red giants as well, should be
examined.

Various binary star systems were included in the PZ program. If
the star-star separation is less than about one-third or greater than
about three times the radius of the habitable zone, then planets in this
zone will have stable orbits for billions of years[7]. Of course, it is
conceivable that planets do not form in multiple star systems.

In their search for narrowband signals, PZ used the NRAO 384 channel
autocorrelation receiver. One hundred and ninety-two channels covered
10 MHz total bandwidth and the other 192 channels covered 625 kHz. The
latter 192 yielded a spectral resolution of 4 kHz per channel, just
narrow enough to discriminate against emission by interstellar hydrogen
atoms. The spectrometer was centered at the λ 21-cm rest wavelength of
the hyperfine (spin-flip) transition of hydrogen in the rest frame of
each observed star. Stars were observed on the order of 4 minutes per
day for approximately 7 successive days. About 10 stars that showed
"glitches" – time variable signals – were reobserved, usually after a
delay of about one year. In only one such case was a second glitch
observed. In no case was the glitch duty cycle large enough to justify
much optimism that an ETI signal had been observed since the main
protection against terrestrial interference is the "protected" nature
of the 21-cm band. The detailed results of the PZ program have yet to
be published.

Search programs of this type will miss leakage signals such as the
TV carrier signals for Mork and Mindy and even the stronger military
radars by many orders of magnitude if they are transmitting at the same

power levels as we now do[8]. However, a narrowband signal from a 40-megawatt transmitter on a 300-ft. antenna could have been detected by PZ as far away as the most distant stars they observed.

Some of the stars examined in the four targeted stellar search programs in Table 1 are sufficiently close that there would have been time for an alien civilization at the star to have detected our leakage radiation and to have beamed an "answer" back to us that could have arrived by 1972 (or later). Two rather solar-like stars, τ Ceti and ε Eridani, are only ∿ 10 light years from the Earth and, therefore, satisfy this constraint. These were the two stars examined at length by Frank Drake in project Ozma in 1960 and recently by PZ and others more briefly, but with much greater sensitivity.

A search project that is fairly similar to the PZ program is being carried out at the Canadian National Radio Observatory in Algonquin Park by Alan Bridle and Paul Feldman (BF). The primary difference is that BF observe at the water line wavelength 1.3 cm (22.2 GHz). Consideration of non-instrumental sources of noise - e.g., atmospheric, galactic, 3 degree cosmic background - suggest that in an optimized ground-based search program wavelengths between 10 and 20 cm will be quietest (e.g., Fig. 5-2 in Reference 1). According to this figure, 1.3 cm is inferior to 21-cm mainly because of absorption of the 1.3 cm waves in the Earth's atmosphere. However, at the present time, many receivers and telescopes are far from optimum; the largest contribution to the system noise temperature is instrumental and Fig. 5.2 is not especially relevant. Thus, the relatively poor sensitivity of the BF program is due more to an inferior receiver and a small telescope than to the effects of the Earth's atmosphere.

BF covered a total bandwidth of 10 MHz with 30 kHz resolution. They have, so far, examined 70 stars within ∿ 45 light years of the Earth during a total of ∿ 140 hours of observations. Results are not yet published. The BF program is presently in cold storage but will be resumed if and when a better 1.3 cm receiver becomes available at Algonquin Park.

The program carried out by Paul Horowitz has been completed and results are published[9] (but not a list of target stars). Because he used a very large telescope and ultrahigh spectral resolution, Horowitz was very sensitive to very narrow signals but he covered only a limited phase space.

Horowitz search a total bandwidth of only 1 kHz but with 0.015 Hz resolution! (65,536 equally spaced frequency bins). This resolution was dictated by two considerations: the short term stability of the Arecibo rubidium reference oscillator and the ultimate limit to narrowband interstellar radio propagation - line broadening by multiple scatterings from fluctuations in the ionized component of the local interstellar medium. Because he was able to cover only 1 kHz of bandwidth, Horowitz adopted a search strategy which assumed that the

transmitting society has accurately measured the radial velocity of our
Sun and/or the velocity of the center of mass (barycenter) of the solar
system - the two differ by at most ± 60 Hz at 21-cm wavelength. Therefore, They transmit at a frequency such that their signal arrives in our
solar system at the hydrogen-line rest frequency of either the solar or
barycentric system. It is interesting to note that within the next few
decades the best radial velocity measurements of solar type stars we
are likely to make are at the 10 m/s level. The accuracy will be
limited most probably by bulk motions in the stellar atmospheres.
Ten m/s corresponds to $\sim$ 50 Hz at 21-cm wavelength.

Because of the spin and orbital motions of the Earth, it was
necessary to update the local oscillator thousands of times during the
few hundred seconds that characterized Horowitz's observations of a
given star. (See Figure 1 in reference 9 for a block diagram.) Data
were recorded on a 9-track magnetic tape, fast Fourier transformed
off line, and displayed in a 256 x 256 raster.

Horowitz examined F, G, and K main sequence stars (from the RGO
catalog), but excluded all known binaries. Total telscope time involved
was 80 hours. A megawatt transmitter on an Arecibo-like antenna could
have been detected out to distances of 1000 parsecs, if the bandwidth
of the signal was < 0.015 Hz. Horowitz showed that one can construct
a 65,000 channel radiometer with very narrow resolution over a limited
frequency range with only moderate computational effort and present the
results in such a way that they can be studied easily by people. In
addition, when one observes with very narrow bandwidths, terrestrial
interference becomes a negligible problem. Horowitz had no false alarms.

Another recent novel search technique, due to Jill Tarter et al.[10],
also achieved excellent frequency resolution and many channels. One-bit
sampled data were recorded on a high speed Very Long Baseline Interferometer tape recorder (recording rate = 720 kb/s). The data tapes recorded
at NRAO were read and analyzed in post-real time, and yielded a total
spectral coverage of $\sim$ 1.2 MHz with 5.5 Hz resolution - the equivalent
of a $\sim$ 200,000 channel spectrum analyzer. (A block diagram is given in
Figure 1 of reference 10.)

During a single observation of a target star, one magnetic tape
was recorded in approximately 4 minutes which is well matched to the
limited tracking capability of the NRAO 91- m transit telescope.
Integration times were sufficiently short that the maximum frequency
drift at 1666 MHz (18 cm wavelength) due to Earth rotation was
comparable to the 5.5 Hz resolution. Therefore, Doppler drift
corrections to update the LO system were not required. Δf can be
reduced below 5.5 Hz in this type of experiment, but then the data
reduction becomes more difficult.

Tarter et al. observed at the upper (OH) end of the "water hole"[1].
Like Horowitz, they examined (apparently) single F, G, and K main-
sequence stars from the RGO catalog.

This program is continuing and promises to yield the best frequency resolution until mega-channel spectrometers are built. Its disadvantages, modest instantaneous bandwidth and large computational overhead, can be partially overcome by the use of dedicated mini-computers and special purpose hardware processors.

All four of the searches discussed above concentrated on nearby stars. An "all-sky" survey is also feasible even though it takes a lot longer to complete. Most of the time only distant stars or galaxies are in the telescope beam and such distant civilizations are unlikely to be directly beaming at our solar system (unless perhaps one of their probes happened to be here!). Therefore, when compared with the targeted search mode, a given power received at the Earth in the all-sky mode would, in general, imply a very much larger power source at the transmitting end.

An all-sky survey is currently underway at the Ohio State University Radio Observatory[11,12]. About half the sky has been searched during an essentially continuous effort begun in December 1973. The spectral coverage (500 kHz total bandwidth at 10 kHz resolution) and sensitivity are only modest but the virtue of this program is its longevity and the opportunity to frequently recheck any interesting directions in the sky. The most interesting signal detected during the first six years of searching is described in reference 12.

It has been argued[13] that the absence of extraterrestrials on the Earth already dooms to failure any search for galactic civilizations. However, if the nearest technical civilization has emerged in a nearby galaxy rather in the Milky Way, then They may not yet have had suffi-cient time to travel here. A search of nearby galaxies could reveal only Kardashev Type II or Type III civilizations that are capable of generating enormously more powerful radio beacons than our own Type I civilization (see Appendix B in reference 1). A brief search ($\sim$ 100 hours of telescope time) of 5 nearby galaxies was carried out by Frank Drake and Carl Sagan at the Arecibo Observatory. Leo I, Leo II and M49 were each covered by observations at 9 positions. Many positions were observed in M31 and M33. Each galaxy was observed at 3 separate wavelengths (21 cm, 18 cm, and 12.5 cm). In all cases, 3 MHz of bandwidth was covered with 1 kHz resolution using the 1008 channel correlator at the observatory. For observations of M33 at the H (21-cm) and OH (18-cm) wavelengths, the spectrometer was centered at the radial velocity of the known hydrogen emission at each position that was observed. Observations of the other galaxies were centered at the systemic velocity where known or else all possible systemic velocities for the local group were used.

The hope was to find one super-civilization in the $\sim 10^8$ stars included in the Arecibo beam during each observation. The results of the Drake/Sagan program have not been published.

What of future searches at comparable levels of sensitivity? The Ohio State Survey and the Tarter search are continuing and the

Bridle/Feldman program may emerge from cold storage if a new 1-cm system is installed on the ARO 46-m telescope. In addition, SETI may parsitize the radio astronomical community in two ways. One is to hop aboard radio observations being carried out for other purposes[14], by constructing automated SETI backend systems. The other takes advantage of the sad fact that, after $\sim$ 20 years, many radio telescopes are considered obsolete for radio astronomy and are consequently under-subscribed. Such telescopes could be dedicated in part or entirely to SETI.

What of future searches at much improved levels of sensitivity? Technology now exists to produce both the mega-channel spectrum analyzers and the sophisticated signal detectors required to analyze the output of these spectrum analyzers in real time. These backend devices and state-of-the-art low noise feeds and receivers could be employed at existing radio telescopes in order to systematically observe the nearest solar type stars and make a complete sky survey over a much broader range of frequencies at much greater sensitivity than has been possible to date. All of the searches listed in Table 1 (and all other searches of which we are aware) have made use of equipment originally designed and constructed for radioastronomical obser-vations. SETI searches with such equipment are never optimum as one is looking for something which conspicuously does not match the type of emission the hardware was designed to detect. Non-optimum, but it is cheap! The question is whether funds should be invested to build SETI-specific hardware in order to increase sensitivity and make more efficient use of observing time on existing telescopes.

Answers to this question will depend upon each individual's assessment of the size of N and, as we have seen at this meeting, there is no concensus with regard to that number. In addition, the answer must depend on the significance of the negative results of the searches conducted to date and the expected increase in significance of the results from any expanded search. Here it may be possible to achieve more agreement than in the matter of N.

Any search for manifestations of technology from extraterrestrial intelligence must explore an eight-dimensional parameter space which Frank Drake has called "The Cosmic Haystack". The eight parameters are: 3-spatial, 1-temporal, 1-frequency, 2-polarizations and 1-trans-mitted power. Figure 1 is an attempt to compress these 8 dimensions into 3 and depict just how large the haystack might be. To do this, it is necessary to assume that 2 orthogonal polarizations are being received simultaneously, that the duty cycle of the signal is high and therefore the probability of detection is independent of time and finally that any modulation present is intended to make the detection problem easier and does not make the signal more noise-like. Two of the three spatial directions can be represented on one axis as the number of targets or the number of telescope beams needed to tessellate the sky. The frequency axis covers the entire microwave region of the spectrum from 300 MHz to 300 GHz, but even this may be an underestimate. The

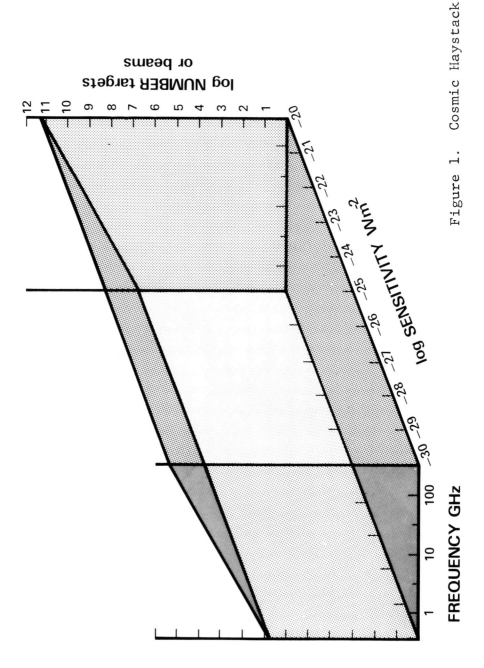

Figure 1. Cosmic Haystack

remaining axis of Figure 1 combines both the third spatial dimension
and the unknown transmitter power; this axis is the sensitivity of the
search measured in Wm^{-2} received within the narrowest channel of whatever detector is being used. The boundaries on this last parameter are
the most arbitrary. The low sensitivity end has been set at 10^{-20} Wm^{-2},
which corresponds to 1 Jansky over 1 MHz of bandwidth and is roughly
the level at which previous radioastronomical surveys of the sky should
have detected a signal if such existed at the frequencies of these
surveys. The high sensitivity limit is what would be required to
detect the planetary radar transmitter at Arecibo Observatory, if it
were located on the far side of the Galaxy. This is roughly the
sensitivity of Cyclops array[1]. The sloping ceiling to this Cosmic
Haystack has been drawn as the number of directions on the sky
(increases as frequency squared) that our largest telescope (Arecibo)
would need to be pointed in to conduct an all-sky survey (assuming of
course that such a telescope could see the whole sky, Arecibo can't).

If each unit on each axis is given equal weight, then there are
some 3×10^{28} cells (of size 1 Hz x 1 Arecibo Beam x 10^{-30} Wm^{-2}) which
may have to be examined in order to complete a systematic search of our
own Galaxy! Clearly not all of these cells are of equal importance.
The observations listed in Table 1 represent attempts to concentrate
on certain more "likely" cells within the constraints imposed by
available astronomical instrumentation. How much of the haystack has
been searched in this manner? Figure 2 is a summary of the programs
in Table 1. It should be immediately obvious that only a small fraction
($\sim 10^{-17}$) of the haystack volume has been explored thus far. In spite
of the fact that considerable effort has been expended by the observers
involved in each of the searches, the significance of their negative
results may not be very great.

Simultaneous sensitivity, bandwidth coverage and frequency
resolution over the entire haystack volume requires a major effort
including construction of dedicated collecting area[1]. However, moderate
coverage of all these parameters (representing a factor of 10^7 improvement over Figure 2) could be achieved in the next decade with existing
antennas and the SETI-specific instrumentation mentioned above.
Figure 3 shows the volume of parameter space that might be covered by
this approach. Studies of cost effective ways of achieving this
coverage are being conducted at NASA-Ames Research Center and the Jet
Propulsion Laboratory. While this approach may prove too grandiose
for SETI skeptics, it is important to remember that a systematic
search for evidence of extraterrestrial technology has not yet been
conducted and is ultimately required to answer the questions: "Where
Are They?", "Are We Alone?". Although some of the observational
programs of Table 1 are continuing, it is probable that their main
contribution will be to provide operational experience in how to
conduct a systematic search.

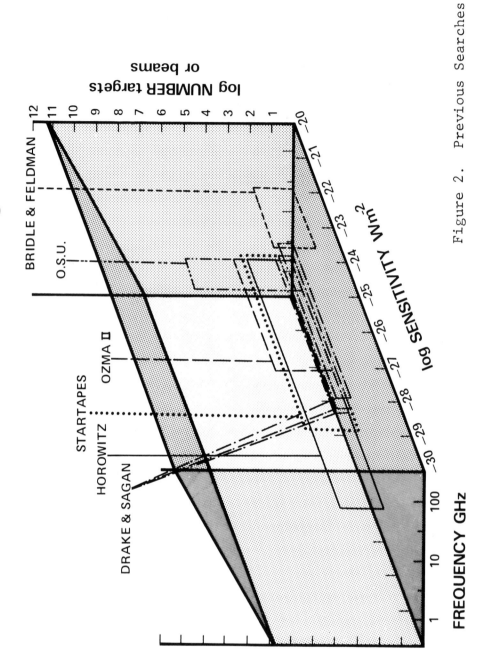

Figure 2.   Previous Searches

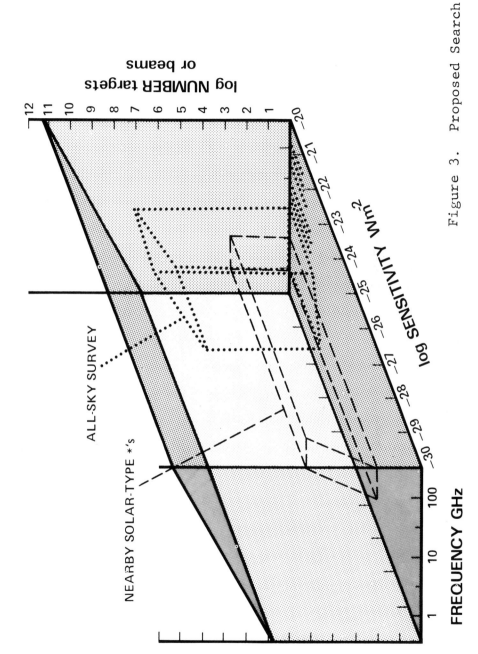

Figure 3. Proposed Search

REFERENCES

1. Oliver, B.M. and Billingham, J.: 1971, Project Cyclops Report,
        NASA CR-114445.
2. Gliese, W.: 1969, Catalogue of Nearby Stars, 1969 Edition,
        Veröffentlichungen Des Astron. Recheninstituts Heidelberg,
        Nr. 22.
3. Wooley, Sir R., Epps, E., Penston, M. and Pocock, S.: 1970,
        Catalogue of Stars within Twenty-Five Parsecs of the Sun,
        Royal Observatory Annals No. 5, Herstmonceux: Royal
        Greenwich Observatory.
4. Sagan, C. and Drake, F.: May 1975, Scientific American, p. 80.
5. Kuiper, T.B.H. and Morris, M.: 1977, Science, 196, 616.
6. Hart, M.H.: 1979, Icarus, 37, 351.
7. Harrington, R.S.: 1977, A.J., 82, 753.
8. Sullivan, W.T., Brown, S., Wetherill, C.: 1978, Science, 199, 377.
9. Horowitz, P.: 1978, Science, 201, 733.
10. Tarter, J., Cuzzi, J., Black D. and Clark, T.: 1980, Icarus
        (in press).
11. Dixon, R.S.: 1979, private communication.
12. Kraus, J.: 1979, Cosmic Search, 1, 31.
13. Hart, M.H.: 1975, Q.J.R.A.S., 16, 128.
14. Langley, D.: 1976, unpublished MS thesis from Electrical
        Engineering and Computer Science Department, University of
        California, Berkeley. Also breif description of SERENDIP
        system appears in Murray, B., Gulkis, S. and Edelson, R.:
        1978, Science, 199, 485.

A BIMODAL SEARCH STRATEGY FOR SETI

Samuel Gulkis and Edward T. Olsen
Jet Propulsion Laboratory
California Institute of Technology

Jill Tarter
University of California, Berkeley

INTRODUCTION

Thus far in this meeting we have discussed some theoretical aspects of the likelihood of intelligent life appearing in our galaxy and
have reviewed the search programs which have been carried out to date.
It is evident from these discussions, as it was twenty years ago when
Coconni and Morrison (1959) first published their classic paper on the
use and detectability of the 21 cm wavelength for SETI, that the detection of an artificial signal of extraterrestrial origin will not be any
easy matter. Whether N is large or small, we do not know where to point
our antennas, on what frequency to listen, nor the type of signal to
search for. To determine these, we must engage in an extensive and
systematic search program.

About four years ago, groups at the NASA Ames Research Center (ARC)
in Mountain View, California and at the Jet Propulsion Laboratory (JPL)
in Pasadena, California set out jointly to develop a search strategy
which had as its objective to greatly expand the parameter space which
has been observed to date, and to accomplish this with existing telescopes. The search strategy developed by these groups is based on
observations to be carried out primarily in the terrestrial microwave
window (defined here to be the spectral range from 600 MHz to 25 GHz).
This is the spectral region in which the sky brightness is minimum,
which contains the "water hole" and microwave water line frequencies,
and for which instrumentation is readily available.

The search strategy assumes that signals of extraterrestrial origin may be narrowband transmissions which are continuously present or
pulsed. The proposed strategy will be most successful at detecting
those types of signals which are designed to be easily detectable but
will nevertheless retain some sensitivity to more complex types of
signals. The observational program embodied in the ARC/JPL plan is
bimodal in character, to cover a wide range of possibilities (Seeger,
1979). One goal of the program is to survey the entire sky over a wide
range of frequency to a relatively constant flux level (Murray et al.,

93

M. D. Papagiannis (ed.), Strategies for the Search for Life in the Universe, 93–105.
Copyright © 1980 by D. Reidel Publishing Company.

1978). This survey ensures that all potential life sites are observed to some limiting equivalent isotropic radiated power (EIRP) depending upon their distance. A second goal is to survey a set of potential transmission sites selected a priori to be especially promising, achieving very high sensitivity over a smaller range of frequency (Oliver and Billingham, 1972). The purpose of this paper will be to discuss the various aspects of these two complementary observational goals embodied in the ARC/JPL search strategy. Table 1 summarizes the principal observing parameters of these two approaches.

TABLE 1

	ALL SKY SURVEY	SELECTED SITE SURVEY
Coverage	4π Steradians	$\sim$800 beam areas
Sensitivity Limit	$\sim 10^{-23} \sqrt{f_{GHz}}$ W/m^2	10^{-25} –10^{-27} W/m^2
Frequency Range	1 – 10 GHz + Spot Bands	1 – 3 GHz + Spot Bands
Spectral Resolution	32 Hz	1 Hz
Integration Time/Beam Area	0.3 – 3 sec.	100 – 1000 sec.
Total Integration Time for Survey	1.6 yrs.	$\sim$0.5 yrs.
Aperture	34–m	$\geqslant$64–m

SENSITIVITY LIMITS FOR EXISTING TELESCOPES

It is presumed at the onset of the ARC/JPL program that existing radio telescopes, equipped with state-of-the-art receivers and data processing devices, will have both the sensitivity to explore the vicinity of nearby stars for transmitters similar to earth's, and to explore the entire galaxy for more powerful signals or for signals beamed at us. In this section, we examine this assumption on a quantitative basis.

An isotropic transmitter with power P_T at a distance r will create a flux

$$\Phi = P_T/4\pi r^2 . \tag{1}$$

The power received by an antenna of physical area A and aperture efficiency η, from a source which produces a flux, ϕ, is given by

$$P = \phi \, \eta \, A. \tag{2}$$

This power can be detected when it exceeds the inherent noise fluctuations of the system. For a post-integration, power detection receiver with equivalent input noise temperature T and IF bandwidth b, the standard deviation of the noise fluctuations is given by

$$\sigma = k \, T \sqrt{\frac{b}{t}} \tag{3}$$

where k is Boltzman's constant and t is the integration time. Thus, detection of a signal is possible when the signal power, P, exceeds the noise power, σ, by some specified factor, α, determined by the allowable probability of a false alarm. Hence, we have the condition for detectability given by

$$P \gtrsim \alpha \sigma = \alpha \, k \, T \sqrt{\frac{b}{t}} . \tag{4}$$

Combining Equations (2) and (4), we can write for the minimum detectable flux (in W/m^2)

$$\phi = \frac{\alpha \, k \, T}{\eta \, A} \sqrt{\frac{b}{t}} \tag{5}$$

Table 2 gives sensitivity limits for some existing telescopes instrumented with low noise receivers. The parameters used are as follows:

$$\alpha = 5$$
$$b = 1 \text{ Hz}$$
$$\tau = 1 \text{ s and } 1000 \text{ s}$$
$$\eta = 0.5$$

If 1000 seconds is the integration time on each star, sensitivities in the range 4.9×10^{-27} W/m^2 to 3.3×10^{-25} W/m^2 are achievable from the largest to smallest telescope for spectral resolution of 1 Hz. The combined frequency resolution and sensitivity (1/2 jansky in a 1 Hz bandwidth for Arecibo) fall far outside that which has been used in radio astronomy.

From Equation (1) the flux received at the earth from a transmitter r light-years away, per watt of equivalent isotropic radiated power is

$$\phi = \frac{8.89 \times 10^{-34} \text{ EIRP}}{r^2} \quad \text{W/m}^2. \tag{6}$$

We may write for the range of a detectable signal

$$r = \left[\frac{8.89 \times 10^{-34} \text{ (EIRP) } (\eta) \text{ (A)}}{\alpha \, k \, T} \sqrt{\frac{t}{b}} \right]^{1/2} . \tag{7}$$

TABLE 2

ϕ (W/m^2)

TELESCOPE	T_s(K)	A (m^2)	τ = 1 sec	τ = 1000 sec
Arecibo	40	35633	1.55 x 10^{-25}	4.89 x 10^{-27}
Greenbank 93-m	30	6808	6.07 x 10^{-25}	1.92 x 10^{-26}
Parkes or DSN 64-m	20	3217	8.58 x 10^{-25}	2.71 x 10^{-26}
Ohio State	40	2221	2.48 x 10^{-24}	7.86 x 10^{-26}
Stanford 5-element Interferometer	20	1313	2.10 x 10^{-24}	6.65 x 10^{-26}
DSN 34-m	20	908	3.04 x 10^{-24}	9.61 x 10^{-26}
Greenbank or DSN 26-m	20	527	5.24 x 10^{-24}	1.66 x 10^{-25}
Stanford 18-m	20	263	1.05 x 10^{-23}	3.32 x 10^{-25}

Figure 1 shows the minimum detectable EIRP as a function of distance
for a typical sky survey (α = 5, b = 32 Hz, τ = 1 sec) carried out
with a 34 m telescope, and for two telescopes used in a discrete source
survey (α = 5, b = 1 Hz, τ = 1000 sec). We see from this figure that
the Arecibo telescope can detect 10^8 W (EIRP) out to about 4 light-
years, the distance to the nearest stars. Transmitters at the level of
the strongest TV stations, 10^5 - 10^7 W (EIRP), can be detected only if
they are situated less than 2 light-years away. A transmitter equivalent to the most powerful radar systems used on earth, 10^{13} W (EIRP),
can be detected by the sky survey at a distance of 22 light-years and
by the discrete source survey at a distance of 1348 light years.

BIMODAL OBSERVATIONAL PLAN

Discrete Source Mode

The a priori site survey is designed to observe 773 stars which
have been identified within 25 parsecs of the sun to be of spectral
type F, G, or K and luminosity Class V. The frequency range to be
covered will be 1.2 GHz to 3 GHz and as many spot bands between 3 GHz
and 25 GHz as time permits. This spectral region includes the water
hole, 1.4 GHz to 1.7 GHz, which has been suggested as a preferred frequency band for an interstellar search (Oliver, 1979).

The sensitivity level which can be achieved depends on both the
telescope used and the integration time. Table 2 gives typical sensitivities that can be achieved with the Arecibo antenna and with several
smaller DSN antennas. Only 244 stars are visible to the Arecibo telescope; the other 529 stars must be observed using telescopes more than
a factor of 3 smaller in diameter. The integration time required to
achieve a constant sensitivity varies as the inverse fourth power of
the telescope diameter. Thus, no attempt will be made to achieve a
constant sensitivity search as this would require $\sim$260 times as much
observing time on a 64 m antenna as on the Arecibo antenna. Rather,
all stars shall be observed for approximately the same integration
duration.

Sky Survey

The sky survey will search the entire celestial sphere over the
frequency range 1.2 < f < 10 GHz inclusive and as many spot bands between 10 GHz and 25 GHz as time permits. In order to better understand
the constraints on time and sensitivity of a sky survey, it is convenient to modify Equation (5) by substituting for the integration time,
t, the amount of time required to sweep the telescope primary beam past
a location on the celestial sphere. Let ω be the angular tracking rate
in deg/second. Then the time required to move a half power beamwidth is

$$\xi = 70 \ c/f \ D \ \omega \qquad\qquad (8)$$

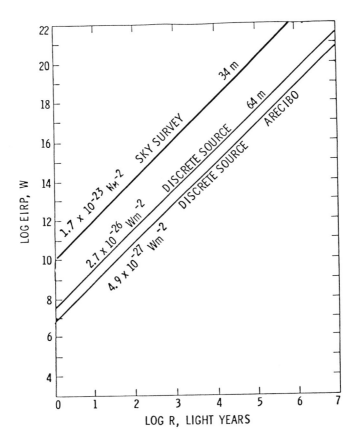

Figure 1.   The minimum equivalent isotropic radiated power detectable
            as a function of distance from the transmitter for a signal-
            to-noise ratio equal to unity. The all sky survey is as-
            sumed to be carried out at a spectral resolution of 32 Hz
            and a one-second integration period on a 34-m telescope.
            The discrete source survey is assumed to be carried out at
            a spectral resolution of 1 Hz and a 1000-second integration
            period on a 64-m telescope and the Arecibo telescope.

in seconds, where D is the diameter of the telescope, f is the observation frequency, and c is the velocity of light. Since $A = \pi D^2/4$ for a circular aperture, Equation (5) becomes:

$$\phi = (4 \alpha k T_s/\pi \eta) \sqrt{\omega b f/70 c D^3} \tag{9}$$

in W/m^2.

Now the time required to survey a fraction, g, of the sky is

$$T_i = 4 \pi g \xi/\Omega_i \tag{10}$$

where the solid angle of the beam is

$$\Omega_i = \varepsilon c^2 /\eta A f_i^2 \tag{11}$$

and ε is the beam efficiency. Substituting (11) into (10) and assuming a circular aperture, the time to survey at a particular frequency setting is found:

$$T_i = (70 g \pi^2 \eta D f_i/(c \omega \varepsilon). \tag{12}$$

The time to carry out a complete survey between the frequency limits f_L and f_u depends on both the bandpass, B, of the multichannel spectrum analyzer and the scan rate. The bandpass B is equal to the product of the number of channels in the spectrum analyzer and the single channel bandwidth (b). An attractive operational procedure is to operate at a constant angular tracking rate. This choice results in a sensitivity which varies as the square root of the frequency. This leads to the following expression for the time to carry out a survey between the frequency limits f_L and f_u

$$T = [(70 g \pi^2 \eta D)/(2 c \omega B \varepsilon)] (f_u^2 - f_L^2). \tag{13}$$

Assuming that the survey is carried out using a 34 m telescope which scans at a rate of 0.2 °/sec, the time required to survey the entire sky at a single frequency setting is

$$T_i = 3 \sqrt{f_{GHz}} \text{ days} \tag{14}$$

if the scans are separated by one half-power beam width. The instantaneous bandpass of a 2^{23} channel analyzer operating at a spectral resolution of 32 Hz is 268.4 MHz. Thus the time required to survey the entire celestial sphere over the frequency range $1.2 \leq f \leq 10$ GHz is

$$T \approx 1.6 \text{ years.} \tag{15}$$

The sensitivity achieved in such a survey shall be

$$\phi = 2.0 \times 10^{-24} \alpha \sqrt{f_{GHz}} \text{ W/m}^2. \tag{16}$$

Instrumentation

        Figure 2 shows a block diagram of a system instrumented for SETI
observations. The complete system consists of the radio telescope,
orthogonally polarized feeds, low noise receivers, a large multichannel
spectrum analyzer (MCSA) and associated accumulators, processor, and
various recording devices.

        The crucial new instrumentation which enables this search to
greatly expand on previous searches is the large multichannel spectral
analyzer. Such devices will make it possible to examine large numbers
of frequencies simultaneously. A digital spectrum analyzer with 300 MHz
of bandwidth and a million channels is currently under construction at
the Jet Propulsion Laboratory (Morris and Wilck, 1978). This device
produces FFT transforms in 10 MHz sections with 300 Hz resolution. One
such device designed at Stanford University in collaboration with the
Ames Research Center and JPL yields 10^7 channels over 8.4 MHz with a
resolution of 1 Hz or 10^7 channels over 268 MHz with a resolution of
32 Hz.

        Figures 3 and 4 show schematic block diagrams of the sky survey
and discrete source survey devices, which consists of two stages of
digital bandpass filters followed by digital FFT (Fast Fourier Transform) processors. The primary spectral resolutions originating in this
design are 8.4 MHz, 65.5 KHz, 1 KHz, 32 Hz and 1 Hz. Synthetic spectral resolutions may be generated by combining samples in frequency and
in time, and the MCSA will be designed to output power spectra over the
range of spectral resolution from 1/4 Hz to 4 KHz in steps which are
multiples of two. This resolution tree allows a more nearly optimum
match to a wide range of continuous and pulsed signal bandwidths. The
bandpass filter design ensure spectral isolation to the 1 KHz level
(256 Hz counting synthetic frequencies), i.e., a strong interfering
signal will not "splatter" more than this amount beyond its own bandwidth. The use of FFT processors at the high resolution end of the
chain minimizes memory required in the hardware. The MCSA architecture
is highly multiplexed so that the actual board level parts count is
minimal.

        The MCSA is designed to be microprogrammable, which allows the
characteristics of the bandpass filters and the hardware thresholding
levels (α) to be changed through software commands; in addition the
nature of the transform performed in the fast processors may be altered
by software commands. The test as to whether the power appearing in a
bin exceeds a preset threshold will be carried out in the MCSA at high
speed. The data corresponding to threshold crossings will be passed
over to the decision processor for further tests.

        The MCSA and decision processor shall be sensitive to three dis-
tinct classes of signals: (1) a continually present, nondrifting signal
of the order of the mean noise level, (2) a periodically pulsed signal
of the order of ten times the mean noise level and which may be drifting,

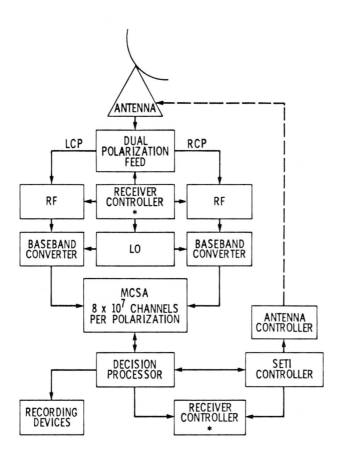

Figure 2.   A block diagram of a system instrumented to carry out
a search for extra terrestrial intelligence. The system
must be capable of processing data on-line to discard
data of no archival value.

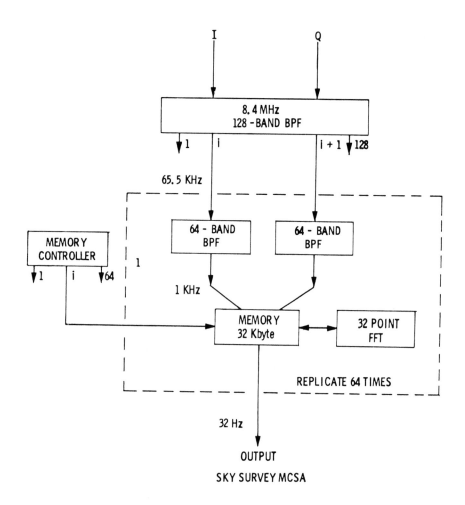

Figure 3.   A schematic block diagram of the multichannel spectrum
            analyzer designed for the all-sky survey. It is designed
            to cover a 268.4 MHz instantaneous bandpass at 32 Hz
            resolution.

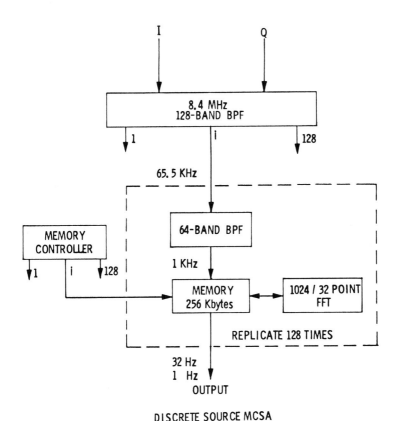

DISCRETE SOURCE MCSA

Figure 4.   A schematic block diagram of the multichannel spectrum
            analyzer designed for the discrete source survey. It is
            designed to cover a 8.4 MHz instantaneous bandpass at
            1 Hz resolution.

and (3) a continually present, drifting signal of the order of the mean
noise level.

Class (1) signals are the easiest to detect, requiring only a
threshold test on the accumulated spectrum at the end of an integration
period. Class (2) signals are more demanding on the hardware/software.
Their detection requires a threshold test for each sample. Data exceeding thresholds is stored over a period of observation and then the
decision processor must examine the resulting matrix for periodicity
and drift. The sensitivity achieved will depend upon the number of
false alarms which the system can handle. Class (3) signals require
large, high speed memory modules, for each sample spectrum must be
stored over some chosen period. Sensitivity is enhanced through removal
of frequency drift before accumulation and thresholding. Of course,
the processor must perform this manipulation over a range of drifts.

SUMMARY REMARKS

It is presently believed that a number of large spectrum analyzers
could be built and used at existing radio observatories to carry out a
significant SETI program. While this program is still in its conceptual
stages, we believe that a field test program could begin as early as
1980-81 using a smaller version of the large spectrum analyzers and
other more traditional hardware. The immediate goals are to learn more
about the radio frequency background (interference) at high spectral
resolution and to develop signal detection algorithms which efficiently
recognize the presence of a signal and automatically attempt to identify
and classify it. It is anticipated that full scale instrumentation
development could begin in 1982 and an operational system placed in the
field in 1984.

ACKNOWLEDGEMENT

We thank our colleagues Dr. B. Oliver, Dr. A. Peterson, and members of the SETI groups at the Ames Research Center and the Jet Propulsion Laboratory for their many contributions to the ideas reflected in
this paper. This paper presents one phase of research carried out at
the Jet Propulsion Laboratory, California Institute of Technology,
under Contract Number NAS 7-100, sponsored by the National Aeronautics
and Space Administration.

BIBLIOGRAPHY

Cocconi, G., and Morrison, P., "Searching for Interstellar Communica-
        tions," Nature, 184, pp. 844 (1959).

Morris, G. A., and Wilck, H. C., "JPL 2^{20} Channel 300 MHz Bandwidth
        Digital Spectrum Analyzer," DSN Prog. Rep. 42-46, pp. 57-61 (1978).

Murray, B., Gulkis, S., and Edelson, R. E., "Extraterrestrial Intelligence: An Observational Approach," Science, 199, pp. 485-492, (1978).

Oliver, B. M., and Billingham, J., "Project Cyclops," NASA CR114445 (1972).

Oliver, B. M., "Rationale for the Water Hole," Acta Astronautica, 6, no. 1-2, pp. 71-79 (1979).

Seeger, C. L., "Strategic Considerations in SETI, and a Microwave Approach," Acta Astronautica, 6, no. 1-2, pp. 105-127 (1979).

P A R T III

THE SEARCH FOR PLANETS AND EARLY LIFE

IN OTHER SOLAR SYSTEMS

SEARCH FOR PLANETS AND EARLY LIFE IN OTHER SOLAR SYSTEMS
AN INTRODUCTION

Jesse L. Greenstein
California Institute of Technology
Pasadena, CA 91125, U.S.A.

The subject discussed by most authors in this section is the potential of our technology to detect planets around other stars. The
subject is of fundamental importance to the question of life, because
we assume that life may originate and evolve only on planets. If
planets were found near other stars, then with the development of proper instruments we could go one step further and search for a spectroscopic evidence of early life on these planets as discussed by Dr.
Owen. Similar spectroscopic studies are currently carried out for the
other bodies of our solar system. As was pointed out by Dr. Soffen in
Montreal, in spite of the generally negative results of the Viking
experiments on Mars, and the rather pessimistic prospects for other
planets and satellites, the possibility that life might exist in our
solar system, outside the earth, has not yet been completely eliminated. Continued exploration of our solar system in the coming decades
could yield useful insights into planetary evolution, the conditions
necessary for the origin of life, and possibly on the question of interstellar travel and galactic colonization as discussed by Dr. Papagiannis. Space exploration, therefore, will remain an important companion to astronomical searches for signs of life in the Cosmos.

As a result of two NASA-backed workshops on planetary detection,
fairly sophisticated thinking has defined some major technical possibilities for direct observation of planets. Two examples are infrared
interferometry (using planetary thermal radiation) or a giant, apodized, low-scattering telescope in space. Two indirect detection methods, from Earth, are based on the induced gravitational wobble of a
star as observed astrometrically, or spectroscopically. Observing
nearby, low-mass stars, the astrometric method at one institution has
been used for over 25 years to find 20 invisible astrometric stellar
companions. But there is disagreement as to whether it has conclusively proved that any object of planetary mass exists. Therefore,
enormous improvements in astrometric techniques are needed, plus the
time to carry them out. The techniques are being tested and will provide important spinoff in our knowledge of low-mass stars, as well as
planets. The radial-velocity technique seems capable of great

M. D. Papagiannis (ed.), Strategies for the Search for Life in the Universe, 109–110.
Copyright © 1980 by D. Reidel Publishing Company.

improvement based on already tested methods. It requires a dedicated
telescope of modest size, and an ultra-stable spectrograph, and similarly requires long times, of the order of planetary periods, i.e.,
years. The limitation here is the intrinsic stability of the velocity
of a star, to levels of a meter per second. In addition to whatever
planets this method detects, it will enormously increase our knowledge
of long or short-term stellar stability, an astrophysically exciting
question. The astrometric method gives direct information, the radial-
velocity method statistical evidence.

Have recent developments in astrophysics cast any future-pointing
shadow on the question of the existence of life, or of planets? Chemistry was only a trivial part of astrophysics until the discovery of
complex molecules in dense interstellar clouds, (from high-frequency
radio astronomy) and of dust grains, carbonates and silicates, surrounding cool stars at their birth and during their late red giant
evolution (from infrared observations). Both techniques suggest that
complex abiotic molecules may be associated with star birth; we do not
know how many of these survive the turning-on of the star's nuclear
energy, or the accretion process onto even small proto-planets. In
fact, in our Solar System, the evidence for the early presence of unstable isotopes, notably $A\ell^{26}$, suggests a period of destructive, radio-
active heating. However, the common presence in dense clouds of dust
grains and "smog" (since most interstellar molecules are poisons or
smell bad, e.g., alcohol, aldehyde, cyanide, hydrogen sulfide) is a
definite plus for expectation that solid bodies exist commonly near
stars. One argument, however, may have lost its early persuasiveness.
Based on conservation of angular momentum, a shrinking star instead of
spinning up was supposed to transfer most of its angular momentum to
a planetary system. We find, now, that mass loss is universal during
star formation. Loss of material carrying high specific angular momentum is seen in the form of stellar winds that may blow to infinity,
even in stars that should be convectively stable. This same loss of
angular momentum continues after the star is formed, i.e., a spindown
phenomenon is common - and has occurred in our Solar System.

Since chemistry is observed to occur in space, and nucleation
builds solids near stars, it is possible that eventually, theory may
predict a size-distribution of solid objects, from dust to planets.
It should predict that at least one Solar System should exist, even if
that one is our own. The scientific interest in such topics, related
to the early history of star formation (from infrared, space and radio
observations) has in the last years given respectability and content
to an otherwise speculative topic. In a decade or two we should hope
to know whether solid platforms for life exist elsewhere. Perhaps radio signals will be detected and bypass our skepticism. This is possible without man travelling through space, but we should not forget that
instrumented interstellar missions to the nearest stars are not beyond
the bounds of reason (although at present possibly beyond the budget!).

THE ASTROMETRIC SEARCH FOR NEIGHBORING PLANETARY SYSTEMS

George Gatewood, John Stein, Lee Breakiron*, Ronald Goebel,
Steven Kipp, and Jane Russell
Allegheny Observatory, University of Pittsburgh,
Observatory Station, Pittsburgh, Pennsylvania   15214

SUMMARY

Extensive testing suggests that astrometric techniques can be used
to detect and study virtually any planetary system that may exist within 40 light years (12.5 parsec) of the Sun. Three years ago the astrometric group at the Allegheny Observatory began an intensive survey of
20 nearby stars to detect the nonlinear variations in their motion that
planetary systems would induce. Several tests conducted to further our
understanding of the limitations of this survey indicated that the photographic detector itself is responsible for most of the random error. A
new photoelectric detector has been designed and a simplified prototype
of it successfully tested. The new detector is expected to be able to
utilize virtually all of the astrometric information transmitted through
Earth's atmosphere. This is sufficient to determine relative positions
to within an accuracy of 2 mas/hr. Such precisions exceed the design
capabilities of the best existing astrometric telescopes; thus a feasibility study has been conducted for the design of an improved instrument.
The study concludes that a new ground-based telescope and a new detector
combined should be able to study stars as faint as 17th magnitude with
an annual accuracy of a few tenths of a milliarcsecond. However, to
obtain the ultimate accuracy possible from current technology, we must
place an astrometric system above Earth's atmosphere. A spaceborne
instrument utilizing the new detector would in theory have sufficient
accuracy to detect any Earth-like planet orbiting any of the several
hundred stars nearest the Sun.

INTRODUCTION

Approximately three years ago, the astrometric group at the
Allegheny Observatory undertook a concentrated study of 20 neighboring
stars in an effort, using current techniques, to detect the perturbations extrasolar planets might introduce into the apparent motions of
their primary stars. This review details the initial results of that

*Now at the Van Vleck Observatory, Wesleyan University

M. D. Papagiannis (ed.), Strategies for the Search for Life in the Universe, 111–154.
Copyright © 1980 by D. Reidel Publishing Company.

and several studies that grew quite naturally from our analysis of its
sources of error. Besides some preliminary results from the survey,
which is still too immature to provide definitive results, we relate
the following: some of our thoughts on the positional precisions required for detection; a brief report on the sources of error in the
photographic program, including a study of the astrometric errors caused
by the atmosphere; a design for a focal plane photoelectric detector
having a more rapid response and greater sensitivity than the photographic plate; a brief report on observational tests with two simple
prototypes of the proposed new detector; a few comments on proposed
interferometric techniques for astrometry; a feasible design for what
we consider to be the optimum ground-based astrometric telescope; and
a design for a spaceborne survey instrument. Finally, we describe what
we consider to be the logical course in applying each of these improvements in a systematic program to detect and study all the planetary
systems within 12 parsec of the Sun.

     Throughout this review we attempt to combine empirical tests and
theory to yield error estimates that include all probable effects.
Past failures to do this have made astrometrists, as a group, wary of
any predictions of extreme accuracy. Oversights are easily made and
their effects are dramatic. Thus we propose step-by-step development
which includes a reassessment of the probable gain of each future phase
with each step. Nevertheless, despite the uncertainty of some of our
most enthusiastic predictions, the trend toward higher precision is
quite clear, and sizable gains in astrometric precision are definitely
indicated. With these will come new insights in virtually every branch
of astronomy and astrophysics and, as the frequency and nature of
planetary systems become known, we will gain a better perspective of
the solar system and our relationship to it.

SURVEY

A. Background

     The distance from which our planetary system could be detected
and studied by an extrasolar observer depends only on the technological
sophistication of the instruments employed and their access to electromagnetic radiation emitted by the Sun since the planetary system evolved.
At least eight planets cause variations in the otherwise nearly linear
motion of the Sun which are greater than the apparent photocentric
shifts that may result from normal sunspot activity. The telltale
perturbations are almost equally evident from any orientation in space
and their detectability can be reduced to parameters in photon statistics, duration of observation, and the field scale of the optical
system used. The astrometric visibility of our planetary system is so
evident that we must first remove the effects of the Jovian planets
from the apparent motion of nearby stars before their position/time
relationship may be studied for evidence of their planets.

Astrometrists have possessed instrumentation and techniques capable of detecting substellar masses for several decades (Schlesinger, 1917). These objects are predicted by theory and suggested by observation to be quite numerous (Abt and Levy, 1976). For a while it seemed that they were evident in many astrometric studies; for example, a veritable menagerie of substellar companions discovered over a period of several decades was listed as recently as five years ago (van de Kamp, 1975) and even more recently by Lippincott (1978). These reports stimulated theoretical papers defining the lower mass of a main sequence star (e.g., Kumar, 1963) and the maximum mass of a planet (Kumar, 1972). However, while confirmed astrometric discoveries have shown the existence of stellar objects whose masses fall considerably below the theoretical nuclear burning cutoff (Heintz, 1978; Lippincott and Hershey, 1972; Gatewood, 1976), they have not shown the existence of objects whose masses are less than the 10 millisun (1 millisun = 1 mS ≈ the mass of Jupiter) degeneracy predicted by Kumar and suggested by him as the upper limit to planetary masses.

This situation may be attributed to the custom in astrometry of modeling the imaging characteristics of an optical system so that the model has the precision of a single plate and then averaging tens or even hundreds of plates into normal points. This leaves the study based on these normal points open to systematic errors nearly as large as the standard error of a single plate. Notably, virtually all of the studies that have reported the discovery of extrasolar planetary masses rely either wholly or partially on observations acquired with the 61-cm visual refractor at the Sproul Observatory. As observations acquired with that telescope have accumulated, it has become increasingly clear that more sophisticated modeling is required. To minimize exposure times, these plates (yellow sensitive) have been exposed only long enough to acquire well-darkened images of the target star and three or four well-placed reference stars. This barely exceeds the minimum requirements for a simple affine transformation of the measured coordinates of each plate into a standard reference frame. The trans-formation produces the constants necessary to place the measured posi-tion into the standard frame with an error only slightly larger than that of the measurement itself. However, the standard frame is defined in terms of magnitude, color, or higher order combinations of the co-ordinates. The first indication of difficulties was noted by Land (1944), who showed that the standard error of normal points formed from observations obtained with that telescope did not decrease with the square root of the number of observations they contained (indicating that they are nonrandom), and that nightly normal points tended to fall in patterns above and below their mean (indicating an unmodeled effect). He also pointed out that the standard error of the positions obtained from a region was dependent on the size of the area over which the reference stars were spread, thus showing that the unmodeled term or terms included parameters in the measured coordinates. Further diffi-culties were indicated when Lippincott (1957) found that a color term was necessary to model the effects of a realignment of the 61-cm lens. Later a distortion term was noted in the region of AC + 65° 6955 by

Hershey (1973). More recently, a cross of coordinate terms into x from y and vice versa was noted in Paper I and in a study combining the observations of the 61-cm refractor with those obtained with the 155-cm astrometric reflector. Strand (1977) showed that the unmodeled color terms of the refractor displace the apparent separation of the red and blue components of Stein 2051 by 37 mas (1 mas = 0.001 arcsec). In the meantime, unsuccessful attempts to confirm the reported planetary systems were underway at other observatories (Gatewood and Eichhorn, 1973; Gatewood, 1974; Gatewood and Russell, 1979). Finally, Heintz (1976) found an unexplainably high number of coincidences in the parameters that describe the small mass orbits suggested by the 61-cm data.

All of these reported difficulties with the Sproul instrument and the failure of other observers to confirm the alleged unseen companions cast grave doubts on the existence of the small-magnitude perturbations observed in that instrument. Unfortunately, due to the lack of well-exposed reference stars, the bulk of these data cannot be rereduced. To include higher order terms in the transformation requires several more reference stars than can be measured on most of the Sproul plates. But even as we discount the planetary discoveries reported so far, several comments must be made. First, to credit telescope errors for the "planets" does not mean those central stars have no planetary systems, only that those planets reported so far do not exist. Also, the Sproul telescope, used with an accurate model of its imaging characteristics and sufficient reference stars, probably has the capability to detect extrasolar planets. Finally, these instrumental difficulties in no way indicate a flaw in the astrometric technique in general.

Instead, these experiences indicate that which, with hindsight, seems obvious. If we are to rely on the randomness of residuals, we must first actively remove their systematic errors. No telescope should be blindly trusted to be free of systematic error at the precision of these normal points. This requires gathering observations by time-delineated groups and processing them as a whole into a well-defined standard coordinate system, testing each group for the presence of significant systematic terms (Eichhorn and Williams, 1963) with the latter sensed against a unified least-squares format. Only in this manner can the accuracy of a normal point be guaranteed to the precision indicated by its statistical weight.

B.   Astrometric Precision and Detectable Mass

Gatewood (1976) has shown that the improved techniques of current astrometry can be used to detect, if present, the Jupiter-like (in mass and orbital period) planets orbiting any of a number of our stellar neighbors. The detectability of extrasolar planets was shown to vary directly with the primary star's parallax and inversely with the two-thirds power of its mass. The arguments presented there indicate that current techniques are sufficiently precise to detect in a 12-year study the existence of a Jupiter-like planet orbiting a star with a detection index of 1.0. A survey with three times this sensitivity could detect, with similar confidence, either a planet with one-third the mass or a similar planet orbiting a star with a detection index of one-third.

These approximations are useful in setting up survey observation programs and can be reasonably carried a bit further. We can, for example, form an approximation for the smallest planet which will probably be detected even for the least favorable orientation of the orbit and instrument. Of course, the smallest mass detectable for the average star in the survey will be smaller than this value. We will call the first of these the "detectable mass" and the second, the "average detectable mass." The detectable mass is a useful concept in the individual study where no planet has been found. Here the investigator is not sure of the orientations involved and must assume the worst when stating that planets of a certain mass and orbital period do not accompany a given star. The average detectable mass applies to a sample large enough to assure a meaningful average orientation.

Presently accepted theory and a number of dynamical considerations (Huang, 1973) indicate that planetary orbits generally have small eccentricities. Little error will be encountered then if we assume that the orbits we seek to detect are circular. A 10% difference in the length of the semimajor and semiminor axes does not occur until we encounter the very unlikely eccentricities greater than 0.45, nearly twice that of any in our system.

As seen from a point in the orbital plane of the planetary system, the star's motion, corrected for all other effects, will appear to be oscillatory along a straight line whose length is twice the orbital radius A. It can be shown that the standard deviation about the mean of a large number of error-free positional observations acquired at random times during several orbits of the system is approximately

$$\sigma_o = A/\sqrt{2} \tag{1}$$

The least detectable position angle for this motion is 45° from either of the axes of the Cartesian coordinate system of the hapless observer. At this angle the perturbation is diminished by $\sin 45^\circ$ being spread into the noise of two sets of measures. From equation (1) we see that the standard deviation of this projected motion (σ_p) on either axis is

$$\sigma_p = A/2 \tag{2}$$

If a large number of observations is acquired with a system for which the standard error of a single observation (σ_1) is well known, we can make use of the approximate relationship:

$$\sigma^2 = \sigma_1^2 + \sigma_p^2 \pm \sigma_1^2 /2 \ (n-4) \tag{3}$$

where σ_p is treated as if it were an independent source of random noise, n is the number of observations, and, as we will see from equation (4), n-4 is the number of degrees of freedom. If the star has been observed for less than several full revolutions of the unseen component, equation (2) will not give a valid approximation of A, but equation (3) may still be used to determine if the effects of an unseen companion have been

noted. In this case, σ will increase with the time span of the observations, eventually becoming significantly larger than σ_1.

A series of positions of a star with linear space motion, reduced
into the standard coordinates (ξ, η), can be modeled by the expression

$$\xi_t = \xi_o + \mu_\xi t + P_\xi \pi + \frac{\Delta\mu_\xi}{2} t^2 \tag{4}$$

where ξ_t is the position observed at time t, ξ_o is the calculated position at some central epoch t_o, μ_ξ is the best linear (proper) motion,
P_ξ is a parallax factor determined by the relative alignments of Earth,
Sun, and the star, π is the annual parallax of the star, and $\Delta\mu_\xi$ is the
rate of change of its proper motion. (The equations in η are, of course,
similar.) The term σ is the standard error of an observation determined from a least-squares solution in which the conditional equation
takes the form of equation (4). If the star is single, we will not be
able to detect a significant difference between σ and σ_1. Likewise if
only a fraction of the orbit is covered by the observations, the standard error of the observation will not be significantly different from
the standard error of measurement. The slope of the perturbation will
be absorbed into μ, while changes in the slope will be absorbed into
$\Delta\mu$; thus at first the perturbations will probably be quite undetectable.
However, if new solutions are run from time to time, as the span of the
observations approaches an orbital period, an increasing amount of the
orbital signal will be noted. In fact the investigator will note that
σ increases until approximately one full period has been observed.
Then slight variations in σ will be noted until several periods have
been observed.

From equations (3) and (4), it can be shown that the number of
observations (n) necessary to detect (at a 68% confidence level) the
presence of the effects of orbital motion on σ is

$$n = 2\sigma_1^2 \Lambda^{-2} + 4 \tag{5}$$

while the detectable mass (m_d) is

$$m_d = 1.4 \ \sigma_1 \ I^{-1} \ P^{-2/3} \ (n-4)^{-1/2} \tag{6}$$

where $I = \pi M^{-2/3}$, the detection index given by Gatewood (1976), and π
is the star's parallax (in arcsec), P is the period (in yr), and M its
mass (in solar masses). The term σ_1 must be an external estimate (in
arcsec), including all sources of random error, and the systematic
accuracy of the observations must be actively assured to the precision
sought in the reductions. Sensible systematic error will invalidate
equations (5) and (6).

In equation (6), n is related to the characteristics of the detector and the apparent magnitude of the target object. In effect the
faster or more efficient the detector and the brighter the target and

the reference stars, the larger n may be. This can be expressed in an
index that will indicate the best choices for a program based on a
specific instrument and detector:

$$K = C \sqrt{t}/I \qquad\qquad (7)$$

where t is integration time (in min) and I is the detection index; C
may be adjusted so that the relative detectable mass (K) equals 1.0 for
some standard object such as Barnard's star - K will then be larger for
more difficult target objects, and smaller for easier target objects.

     In an individual study of a given region, the residuals in equa-
tion (4) are first analyzed for all likely periodicities. Any periods
suggested are then used as starting points in an interative orbital
analysis to determine the most significant orbit near each suggested
period. Next the viability of each set of parameters is appraised by
a test similar to that proposed by Eichhorn and Williams (1963). This
test, in effect, asks if the increased parameter variance associated
with the terms that describe the proposed orbit (or orbits) is greater
than the decrease in residual variance resulting from the added degrees
of freedom in the orbital model. The significance of this magnitude
comparison may then be translated into a t test form. Thus we derive,
over the full range of the observations, a table of all possible orbits
and their likelihood of occurrence. Or perhaps even more importantly,
when no periodicity is considered indicative of a real orbit, we obtain,
with statistical measures of certainty, information on what planets (in
terms of orbital period and detectable mass) do not orbit the observed
star. The latter information may have as much long-term significance as
the discoveries we hope to make.

C.   Current Observing Program

     To achieve high astrometric precision from a photographic program,
we must concentrate our observations on a small number of target objects.
A number of considerations suggest that the maximum concentration is
obtained for target regions spaced approximately 2 hr apart in right
ascension. To select the best target objects for our initial effort,
we calculated the relative detectable mass for a photographic program
for the 300 stars known nearest the Sun. From this, 20 stars in 16
regions were chosen and placed on an intensive program which began in
the fall of 1976. These objects (denoted with an asterisk in table 1),
are observed during the 2 hr period centered on midnight, time not used
in the trigonometric parallax program.

     During its first 3 yr, this survey has achieved an average
precision of 5   mas,   standard error per season per star. This
exceeds the combined average weight of the three-telescope study of
Barnard's star detailed by Gatewood (1976), the most extensive published
astrometric study to date, and reaches the average sensitivity originally
sought. If the observing time was evenly distributed, and the program
continued for 12 yr, the detectable mass for each target star would be

TABLE I – ALLEGHENY OBSERVATORY SURVEY FOR EXTRASOLAR PLANETARY SYSTEMS

No.	Name	RA 1950 h	m	s	Dec °		Spec	π	V	B-V	Mass	I	K	Comments
1*	Groom 34 A	0	15	31	43	44.4	MIV	.296	8.08	1.56	0.30	0.61	5.2	P~2600 yr,SB
2*	Groom 34 B	0	15	34	43	44.7	M6Ve		11.04	1.80	.15	1.03	3.3	ρ=44",FS
3	Van Maanen 2	0	46	31	5	9.2	WDg	.229	12.37	.56	.70	.29		
4*	Tau Ceti	1	41	45	-16	12.0	G8Vp	.277	3.49	.72	.80	.32	9.5	
5	L1159-16	1	57	28	12	50.1	M8 e	.226	12.26	.81	.14	.84		FS
6*	Epsilon Eri	3	30	34	-9	37.6	K2V	.302	3.73	.87	.75	.37	9.2	
7	40 Eri A	4	12	58	-7	43.8	K1V	.205	4.43	.45	.77	.24		
8	40 Eri B	4	13	4	-7	44.1	WDa		9.52	.03	.42	.37		ρA=83"
9	40 Eri C	4	13	4	-7	44.1	M4Ve		11.17	1.64	.18	.64		ρB=7".0,P=250
10*	Stein 2051 A	4	26	48	58	53.6	M4	.179	11.09	1.64	.20	.53	4.1	P=370 yr
11*	Stein 2051 B	4	26	48	58	53.6	WDa		12.44	.32	.70	.23	9.4	ρ=7".2
12*	Ross 47	5	39	14	12	29.3	M5	.165	11.54	1.65	.15	.58	7.3	SD
13	Sirius A	6	42	57	-16	38.8	A1V	.377	-1.46	.00			3.3	P=50yr,ρ=7".5
14*	BD+5°1668	7	24	43	5	22.7	MS	.266	9.82	1.56	.21	.74	2.1	
15	G51-15	8	26	52	26	57.1	M	.281	14.81	2.06	.09	1.38		
16	BD+53°1320	9	10	59	52	54.1	MOV	.166	7.61	1.39	.45	.28		P~700yr
17	BD+53°1321	9	11	1	52	54.1	MOV		7.71	1.36	.44	.29		ρ=17
18*	Groom 1618	10	8	19	49	42.5	K7V	.222	6.59	1.38	.53	.34		FS
19	Wolf 359	10	54	5	7	19.0	M7 e	.423	13.54	2.00	.10	2.00		
20*	Lal 21185	11	0	37	36	18.3	M2V	.386	7.49	1.51	.31	.85		
21*	Ross 128	11	45	9	1	6.0	M5	.301	11.09	1.68	.15	1.06	2.9	
22*	Lal 25372	13	43	12	15	9.7	M3V	.199	8.47	1.47	.34	.41	6.4	
23	BD-11°3759	14	31	35	-12	18.6	M4	.162	11.36	1.63	.19	.48		
24	BD-7°4003	15	16	50	-7	32.4	M5	.153	10.57	1.61	.24	.40		
25*	BD-12°4523	16	27	31	-12	32.3	M5	.249	10.11	1.60	.21	.71	2.1	SB?

TABLE I – ALLEGHENY OBSERVATORY SURVEY FOR EXTRASOLAR PLANETARY SYSTEMS

(continued)

No.	Name	RA 1950	Dec	Spec	π	V	B–V	Mass	I	K	Comments
26*	Barnard's Star	17 55 23	4 33.3	M5V	.548	9.54	1.74	.15	1.90	1.0	P∿450 yr
27*	Struve 2398 A	18 42 13	59 33.3	M4	.284	8.92	1.54	.26	.71	3.7	ρ=16"
28*	Struve 2398 B	18 42 14	59 33.0	M5		9.69	1.59	.21	.79	3.3	
29	G208-44	19 52 16	44 17.4	M	.216	13.41	1.90	.12	.88		
30	G208-45	19 52 17	44 17.4	M		13.99	1.96	.11	.93		
31*	61 Cygni A	21 4 40	38 30.0	K5V	.296	5.23	1.18	.60	.42	2.6	P=650 yr
32*	61 Cygni B	21 4 40	38 30.0	K7V		6.04	1.37	.53	.45	2.4	ρ=24"
33	L 789 - 6	22 35 45	-15 35.5	M6 e	.290	12.18	1.96	.13	.12		
34*	E V Lac	22 44 40	44 4.6	M5 e	.195	10.02	1.38	.24	.51	5.2	FS
35*	Ross 248	23 39 27	43 55.2	M6 e	.312	12.29	1.92	.13	1.24	3.5	FS

*On current photographic survey

e = hydrogen emission, P = period of orbit, ρ = separation or semimajor axis, FS = flare star,
   SD = subdwarf, SB = spectroscopic binary.

0.16 mS times the value of K listed for it in table 1. Unfortunately,
the coverage has varied somewhat more than expected, the brighter stars
at smaller zenith distances acquiring more good exposures than expected
while others received less than hoped. This situation has been partially corrected by a redistribution of observing time and will be
further aided by the addition of an automatic tracker.

Most of the survey stars, 65%, are included on a sizeable number
of early plates in the Allegheny Observatory plate file. Although of
lower concentration and regularity, these observations allow us to better
determine the time-dependent parameters of proper motion and perspective
acceleration, thus lowering the parameter variance of each study and
increasing its sensitivity to small perturbations. This material is
also of value in our attempt to find long-period orbits. From equation
(6) we note that m_d is proportional to the $-2/3$ power of the orbital
period. Thus for a given astrometric sensitivity, the detectable mass
is diminished by over two if its orbital period is three times longer.
By the same arguments, we note from equation (6) that, as a continuing
survey follows the motion of a star long enough for a planet to complete
longer and longer orbits, the detectable mass decreases more rapidly
than the inverse of the span of the observations, or $m_d \propto T^{-7/6}$.

The available plate material has been measured and analyzed for
Barnard's star (Gatewood, 1976), Lalande 21185 (Gatewood, 1974), van
Maanen's star (Gatewood and Russell, 1974), and Epsilon Eridani and
Tau Ceti (Gatewood and Russell, 1979b). The residuals of van Maanen's
star and Tau Ceti are both suggestive of unseen (probably stellar)
companions, but the data are not sufficient for the calculations of
orbits.

Both the sensitivity and program size will be increased significantly if the detector discussed below succeeds. In anticipation of its
success the unmarked stars in table 1 are being added to the current
program. As will be noted below, the photometric characteristics of the
new detector are radically different from those of the photographic plate
and the precision per measurement is nearly constant for stars over a
wide range of apparent magnitudes. Thus the latter additions to table
1 are selected by their detection index alone and, to prevent confusion,
K has been omitted. These stars bring the total to 35 stars in 27
regions, 55% of which have a sizeable collection of early plates in the
Allegheny Observatory plate file.

D. Extending the Survey to a Larger Sample

Our understanding of planetary systems generally depends on our
ability to determine the frequency of occurrence of orbital characteristics and mass distributions of planetary systems orbiting stars of all
spectral types and masses. This task is best accomplished by studying,
at least at first, a homogeneous sample. Stars in the vicinity of the
Sun are, to a good approximation, a valid cross section of those found
throughout the spiral arms of the galaxy. It appeals to reason that,

for any given mass and orbital period, the smallest detectable planetary
mass will be found by directing our study to the nearest stars. This is
quantitatively apparent from equation (6), which may be written as

$$m_d = 1.4 \ \sigma_1 \pi^{-1} \ M^{2/3} \ P^{-2/3} \ (n-4)^{-1/2} \qquad (6a)$$

Thus, to give the maximum range to the masses which can be detected for
each type of star, parallax should be made a prime factor in the selection of the stars to be included in future observation programs.

To obtain a sample of approximately 500 stars, the survey will
have to be extended to parallaxes as small as 80 mas or to a distance
of 40 light years. Within this distance of the Sun there are approximately 10 A, 16 F, 50 G, 80 K, and 330 M type main sequence or near
main sequence (luminosity classes IV, V, and VI) stars. The detection
indices of these stars range from 0.06 to 3.21. Thus current photographic techniques could be employed to survey a small sample of even
the most massive and most distant stars in this group to a detectable
mass as small as 5 or 10 mS.

However, a general understanding of planetary systems will require
a detailed knowledge of several dozen planetary systems to masses as
small as that of Earth and periods shorter than 1 yr. To survey the
large sample necessary to find these planetary systems and to obtain
the extreme precisions necessary to detect and study their smaller
masses will require specially designed, fully dedicated instrumentation
such as discussed below.

PHOTOGRAPHIC TESTING OF SOURCES OF ERROR IN THE SURVEY

A. Random Errors Caused by Photographic Plate

For over a century (Gould, 1866), the photographic plate has
served as the prime detector for long-focus astrometry. Although its
quantum efficiency and resolution are not high by today's standards, it
allows integration of the light intensity in all parts of the field
simultaneously. This latter feature is absolutely necessary for
systems operating beneath Earth's atmosphere in the presence of constant
variations in atmospheric refraction and instrumental tracking. However,
the photographic process has been suspected as a source of both random
and systematic error (Land, 1944; van Altena, 1974; Levinson and Ianna,
1977).

In our examination of the relative importance of the photographic
plate in the formation of the standard error of a position reduced from
the measurement of an average exposure, we adopt the expression

$$\sigma^2 = \sigma_A^2 t^{-1} + \sigma_B^2 \qquad (8)$$

where σ^2 is the variance of the exposure, σ_A^2 is the dependence of σ^2
on the inverse of exposure time t, and σ_B^2 is that portion of σ^2 which

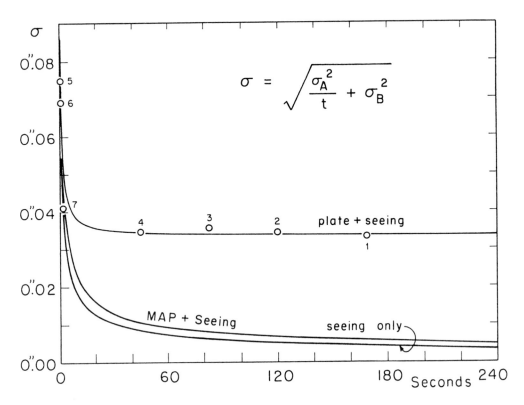

Figure 1 - Points 1 through 4 are the mean standard errors of exposures with integration times t obtained as part of the regular photographic program with the 76-cm refractor. Point 5 was obtained from similar reductions using 1-sec "exposures" obtained from cluster trail plates. Points 6 and 7 are from a short series of 1- and 2-sec exposure plates of the astrometric standard region in Praesepe. The solid line is the best least-squares fit to σ_A and σ_B. The random errors of the photographic process dominate where the exposure time is greater than 4 sec. The random errors of the MAP would enter when those of the seeing are highest. Thus, even if the MAP were no more accurate than the photographic process, 1/2 hr of observation would yield relative positions with a precision approaching 2 mas.

is independent of the integration time; σ_A is essentially the standard error due to the rapid variations in the refractive properties of the atmosphere in the line of sight, while σ_B is the standard error of the detector. That σ_B does not contain significant contributions from slowly varying unmodeled parameters of the Allegheny telescope's optical system and that the -1 power of t is correct to much smaller variances than we are dealing with here is shown later by an intercomparison of the nightly mean positions of field stars.

Points 1 through 4 in figure 1 are each the mean standard error of an average exposure t determined from several hundred plates with t near the indicated means. The value of σ does not show a convincing dependence on t, changing little for exposures as short as 45 sec (point 1). A least-squares reduction of equation (8) to these points yields $\sigma_A = 64 \pm 47$ mas and $\sigma_B = 34 \pm 5$ mas, where t is in seconds. This leaves σ_A poorly established but indicates the importance of σ_B in photographic work.

Two methods were employed to better determine σ_A. The camera was not originally designed to take exposures as short as 1 or 2 sec, so we used a test based on star trails. The telescope was turned to a region of bright stars, a 1-min exposure taken, the shutter closed, and the drive turned off. Once all drive motion had ceased, the shutter was reopened and the star's motion across the focal plane recorded. By measuring the relative north-south positions of the trailed stars, starting at some arbitrary distance from the position indicated by their 1-min exposure and advancing 50 µm along the axis of motion between sets of measurements, we were able to determine the distortions caused by the atmosphere every 0.05 sec of time. These measurements revealed much of interest (KenKnight et al., 1977). In particular, they yielded the standard error of a 1-sec exposure, point 5 on figure 1. Points 6 and 7 result from the reduction of plates taken after alterations to the camera, with 1- and 2-sec exposures of the standard region centered on the Praesepe cluster (Russell, 1976). Utilizing these new data at one-half weight each, a new least-squares reduction for the parameters in equation (8) revealed that $\sigma_A = 61 \pm 6$ mas and $\sigma_B = 33 \pm 4$ mas.

Thus the positional information transmitted by the atmosphere through the Thaw 76-cm photographic refractor at its Pittsburgh site has a potential precision of approximately 1 mas/hr of observation. Under favorable circumstances the current program achieves a normal point with a precision of about 7 mas/hr (standard error). Based on these facts, we are currently detector-limited.

B. Astrometric Errors Caused by Atmosphere

The effects of Earth's atmosphere on astrometric precision have been studied in some detail by a number of authors, for example, Schlesinger (1916), Hudson (1929), KenKnight et al. (1977), and doubtlessly others. Our own studies have added several points of information which will be useful in the present discussion.

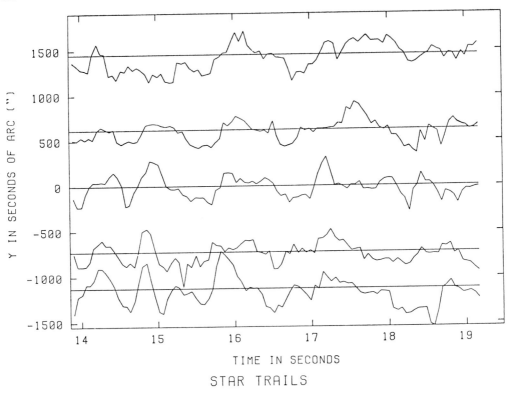

Figure 2 - The above tracings indicate the nature of atmospheric turbulence. Obtained with the telescope drive turned off, the plate records the relative positions of five stars as a function of time. These stars were chosen on the basis of similar apparent magnitude and right ascension. The deviation from their mean positions, determined every 0.05 sec, is magnified 400 times, the plate scale of the telescope being 14.6 arcsec/mm. Preprocessing these data to remove the effects of wavelike patterns could further increase precision of the MAP-seeing curve plotted in figure 1.

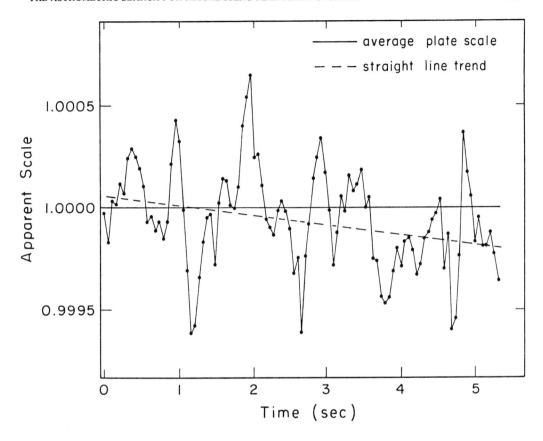

Figure 3 – Variations in the apparent scale of the 76-cm refractor
are plotted as a function of time. The scale term was found from a
trail plate similar to the one illustrated in figure 2. The average
scale of the telescope is shown with a solid line. A trend lasting more
than 5 sec is shown by the dashed line. The scale is given as a
fraction of the normal scale = 14.60 arcsec/mm. Very similar star
trails and atmospheric scaling are found on plates obtained with
several other telescopes.

The multiple-star trail plates discussed earlier allow a study of correlated image motion across the field of the instrument. Schlesinger and Hudson studied the average offset of images from trail plates for intervals of several seconds, finding oscillatory motion with a period of approximately 1 min. This motion was nearly perfectly correlated across their plates. KenKnight et al. found similar patterns of image motion with periods near 1 sec (fig. 2) and near 5 min, with much longer period motions indicated. Besides the oscillatory motion we also found "periodic" variations in the field scale at both periodicities (figs. 3 and 4), which indicate a refractive wave nature for the responsible phenomena and indicate the need for modeling to remove their effects from astrometric positions.

Figure 5 shows the correlation in image motion, measured each 0.05 second, versus field separation. This correlation is related to seeing quality, consistently increasing in periods of bad seeing. This factor has not been removed from figure 5 and is responsible for much of the apparent scatter. Nevertheless, two major features are apparent. There is a large range of separations (from $0.^{\circ}25$ to over 1°) over which the correlation is nearly constant at a coefficient near 0.4, and there is an area of increasing correlation for separations less than $0.^{\circ}25$, approaching 0.9 at separations of several arcseconds.

The standard error of a single trail about its line of motion averages 190 mas per second of time. By averaging the star trail measurements for greater or lesser periods than 1 sec and then finding the standard deviations of these averages, we find that this value varies inversely with the fourth root of the integration period. An inverse-square root relationship in the quantity would indicate random variation about the line of motion. Inspection of figure 2 shows that the motion of the five star trails shown there are highly nonrandom.

The standard error of the difference between the positions of two stars depends on the separation between the stars. For two stars separated by 7.3 arcsec we found this value to be 56 mas per second of time, with the value growing rapidly at greater separations. However, for the close pair, the standard error of the difference varies inversely with the third root of the integration period, indicating that the differences are more nearly random than the absolute motion of either star about its trail. At greater separations the inverse third root character deteriorates toward an inverse fourth root. The reason for the still present nonrandom effect is what we call "scaling," indicated in figure 3 (KenKnight et al., 1977).

The presence of the scale term was found by constants, not unlike plate constants, determined by reducing the instantaneous deflections of several star trails (like those shown in fig. 2) to their known positions in the Praesepe Astrometric Standard Catalog (Russell, 1976). The average 1-sec mean residual from those modeled positions has a standard deviation of 70 mas. This is in good agreement with the

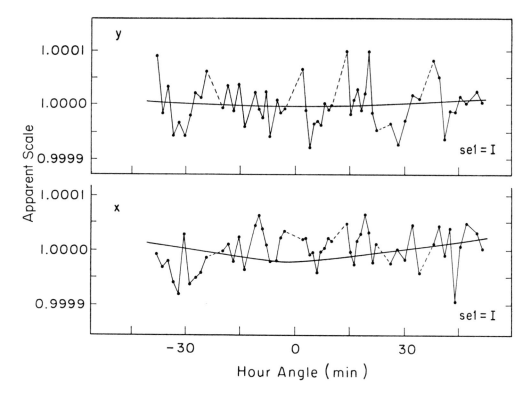

Figure 4.- Atmospheric scaling of a reduced magnitude but increased period is
found by examining the scale terms of successive exposures within a survey
region.   These effects may be modeled and removed, as they were in this
region, if sufficient reference star positions are measured.   The periodicities and magnitudes observed here are similar to those reported for the
Sun (KenKnight et al., 1977).   The refractive features responsible for
some of the longer term variations noted here are thought to be several
kilometers long.

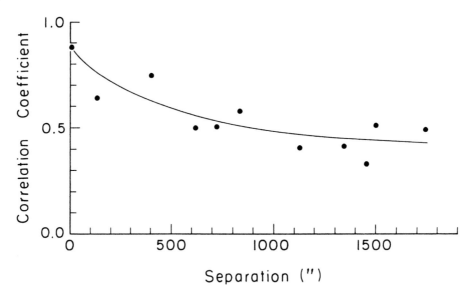

Figure 5 - At separations greater than 1000 arcsec there is little dependence of the correlation coefficient on separation. However, the motion of stars about their mean trails becomes more correlated at smaller separations, reaching 0.89 at a separation of 7.3 arcsec.

average standard error of 1-sec exposures of the standard region which have been measured and reduced by standard techniques (fig. 1). Over the range studied by trail plates, the means of the residuals of these solutions follow an inverse-square-root law in n and appear random. Thus, allowing for both shifts in zero point and scaling greatly increases our ability to predict the position of an image relative to that of its neighbors.

C. Potential Nightly Precision

To test how far the latter effect holds, and thus the ultimate precision obtainable per night from high-speed measurements, modeled as noted, we acquired an average of 42 1-min exposures per night, for several nights, of the same star field. The standard error of each field star position about its mean was then determined. As noted from figure 2, the standard error of a single average 1-min exposure with the Thaw refractor is 34 mas. The average standard deviation of the exposures acquired on each night was found to be 5.4 mas, a value very near that predicted by an inverse-square-root law (t^{-1} in expression (8)). This result suggests that the still unmodeled nonrandom effects in the Thaw telescope's optical system are less than a few milliarcseconds, therefore not affecting σ_B^2 in expression (8), and that the estimated precision of the proposed detector (see next section) combined with the Thaw will on the average be considerably better than the best obtained to date.

D. Modeling Instrumental Projection

If very high precision positions are to be obtained from a given astrometric telescope, it must be possible to model the imaged field of that instrument with an affine transformation (plus any necessary higher order terms) so that the measured field position of an image may be transformed into the relative position of the star. The precision with which this can be done is a function of (1) the errors of the detector and measuring device, (2) the number of relevant parameters in the reduction model, (3) the secular constancy of the relevant parameters, and (4) the number of reference stars measured.

To calibrate various presently and previously used astrometric telescopes to one another, we borrowed and measured 120 plates, from 15 telescopes spanning 113 years of observations, of the region of the Praesepe cluster. To ensure a wide separation of effects, we measured 200 noncluster members, picking those that fell well off the cluster's H-R diagram. Thus the 408 stars measured in the region are practically free of a color-magnitude correlation over their 10 apparent magnitude, 2.0 magnitude color index range. An absolute reference frame was established using the positions of 138 stars which appear in a wide variety of meridian and photographic catalogs.

By careful intercomparison of the positions obtained for these stars from the measurements of each telescope, with the reference frame

positions and with positions obtained by averaging the results of all of
the other telescopes, we were able to derive the parameters of each instrument and often to estimate the consistency of those parameters. For
existent telescopes, the required number of parameters varied from three
to eight. This number is directly related to the number of reference
stars that must be measured to yield an acceptably low parameter variance
(Eichhorn and Williams, 1963). However, assuming an order of consistency
in these parameters similar to that found in the Praesepe study, and
providing that the proper models are used with a sufficient number of
reference stars being measured, we find that high-precision field positions can be obtained with any existing long-focus instrument.

In summary, our photographic tests indicate that: (1) the highly
correlated, atmospherically induced wavelike motion of star images may
be modeled (by parameters similar to those of a standard plate constant
reduction) to precisions of approximately 1 mas/hr, (2) the inverse-
square-root law holds for properly reduced observations to values as
low as 5.4 mas, indicating that the remaining systematic errors of the
Thaw telescopes are less than a few milliarcseconds, and (3) with a
sufficient number of reference stars and proper reduction of systematic
errors, existing astrometric telescopes could achieve precisions well
in excess of those now obtained.

Two conclusions important for the tests outlined later pertain only
to the single star trail. We note that that standard error of 1 sec
averaged positions along a trail is approximately 190 mas, and, if we
take longer averages of the deflections, the standard error decreases
with the inverse fourth root of the integration period.

MULTICHANNEL ASTROMETRIC PHOTOMETER (MAP)

The Multichannel Astrometric Photometer (MAP) is a proposed photoelectric measuring device that meets all the criteria for an astrometric
detector, i.e., it has high spatial resolution and quantum efficiency,
it is free from sensible systematic and random errors, and it provides
concurrent records of a sufficient number of reference stars to allow
optimum use of the optical system. The MAP is two orders of magnitude
more efficient than the photographic plate and will yield, with increased resolution, the real-time relative positions of stars up to 2.5
magnitudes fainter. In essence the MAP converts the measurement of
relative positions to the measurement of phase shifts and electronic
signals. Its success is virtually assured because it requires no complicated or untried technologies.

The MAP was inspired by Drake's suggestion that precise positions
could be obtained by photoelectric timing of stellar transits (Drake,
1975). Consider the brightness measurements obtained with a photomultiplier tube placed behind a Ronchi ruling located in the focal
plane of a telescope. As a ruling line, moving perpendicular to its
length, cuts a star image, the objective appears to darken as in a

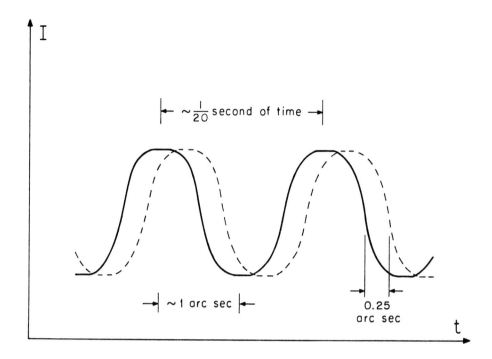

Figure 6 - As a line of the MAP's Ronchi ruling moves across the
star image, the image of the objective appears to darken as in a Foucault
test. However, since the line width is slightly less than the seeing
disk of the image, the image of the objective on the cathode does not
go completely dark. Likewise, the space between the ruling lines is
not sufficient to let all of the star's light through. Thus a series
of maxima and minima are seen (solid line). A second star, viewed by
a second photometer, will cause a very similar output (dashed line).
If the separation between the two stars is known to the nearest whole
number of line spacings, the exact separation in these units may be
obtained by adding to this the phase difference of the two signals.
Illustrated here is the phase difference for two stars 0.25 arcsec
apart. Since only the phase difference is sought, the MAP is insensitive to most of the factors important in absolute photometry.

Foucault test. If the ruling line is slightly smaller than the image, the objective will not go completely dark but will pass through a minimum; if the transparent spaces between the lines have a similar width, the observed illumination will approach a similar maximum. This variation in brightness plotted against time is represented by the solid line in figure 6. If the ruling is long enough and a second photomultiplier tube is placed behind a second image, the observed brightness variations plotted against the same time standard may be similar to those represented by the dashed line in figure 6. Assuming that the approximate distance between the two stars is known to the nearest whole number of ruling lines, the exact distance can be estimated in terms of this number of lines and the phase difference of the patterns (figs. 7(a) and 7(b)).

This principle may be extended to a large number of stars. In essence the MAP modulates the apparent brightness of the reference and target stars against a high-precision, rigid spatial metric, the Ronchi ruling. These brightness variations are amenable to a translation and rotation transformation that reveals the ruling's instantaneous location with respect to its mean on that star field. Fortunately, this transformation, which may be expanded to allow for linear and second-order scale variations as well as tilt of the ruling out of the focal plane, is identical in form to the terms required to gnomonically map the apparent instantaneous positions of the field stars against their mean positions. Thus the errors of the measuring device are absorbed into the star field reduction (similar to plate constants) without an increase in the parameter variance.

The MAP's ability to yield a virtually constant flow of positional information about the target star and a moderate number of reference stars sets it apart from the several other photoelectric detectors that have been proposed. Two of these, proposed by van Altena (1974) and Connes (1978), would determine the positions of the target stars with respect to the average positions of a large number of reference stars, thus losing the individual reference star positions and along with them the ability to model the imaging characteristics of the optical system and passing refractive anomalies. The AMAS (Frederick et al., 1974) would provide the individual star positions, but the need to deconvolve these from the complex signal obtained by viewing the entire field with a single photometer through a rotating Ronchi ruling suggests that the astrometrist will have to be content with the positions of the several brightest stars in the field. In contrast, the MAP takes full advantage of high-speed computer and photoelectric technology to yield all of the dimensional information available for each and every star selected by the astrometrist (fig. 8).

Allowing for the effects of photon noise, Drake has developed an expression for the accuracy with which the centroid of the photocell response can be located. The expression for positional error ε may be shown to be

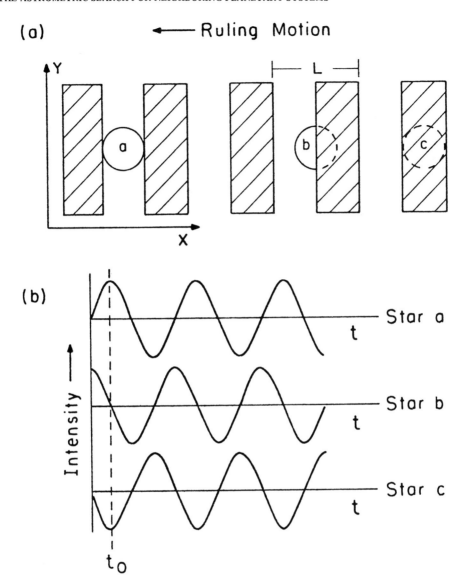

Figure 7 - (a) Ronchi ruling position with respect to star images a, b, and c at time t_o. (b) Signals in the channels monitoring star images a, b, and c showing the relative phases at time t_o.

$$\varepsilon = \frac{\lambda\omega}{D^2 \sqrt{\pi m E I \tau}} \qquad (9)$$

where λ is the central wavelength of observation, ω is the ratio of
image size to the diffraction-limited size, D is the objective diameter,
m is the time of observation during which half or more of the star's
radiation is passed by the ruling, E is the quantum efficiency of the
detector, I is the photon flux in the wavelength band observed, and τ
is the integration period.

Using expression (9) for a 15th magnitude star measured with the
76-cm Thaw refractor using an 800-Å bandpass, ω = 5, E = 23, we find
ε = 1.0 mas on each axis during 10 min of observation. The effects of
atmospheric seeing during this period, that required to secure a positional accuracy for this star of 34 mas photographically, is 3.5 mas on
each axis. Thus, even at the 15th magnitude, photon statistics are not
an important error source. If we assume that the ruling lines are each
known to within 2 μm for a temperature-exposed ruling, the MAP-Thaw
combination should develop a precision of approximately 4 mas within 10
min of observation. Since the Thaw is exposed to all effects of its
environment, it is doubtful that an annual precision exceeding 1 mas can
be achieved, but this would be a factor of 3 improvement over the best
presently attainable.

An added advantage of the MAP is that it will be possible to improve our knowledge of the exact spacings of the lines of the ruling as
the MAP is used. Each measurement of the distance between two stars
yields the relative positions of the lines with which the measurements
were made. Actually, each centroid determination with a 12-channel MAP
would yield the relative positions of each line with respect to the 11
other lines used for that measurement. If we make the ruling more than
twice the size of the field to be measured, we can determine with great
redundancy the location of every line on it. A measuring session with
the MAP, which would consist of a double sweep (forward and reverse) of
the field in each of 4 ruling positions (90° apart), would provide nearly
10^5 centroid determinations and 10^6 intercomparisons of line positions.
After a few years of usage, the relative positions of each ruling line
will be known to a precision limited only by the stability of the material
on which the ruling has been formed.

The MAP may also be used for the real-time calibration of the focal
plane of the telescope. Because it measures the relative positions of
all stars simultaneously, it will be possible to let them drift slowly
about the field. This will require following the motion with the ap-
paratus carrying the light pickups, but not with high precision; a half-
millimeter error should go unnoticed. As the field drifts, the instantaneous nongnomonic characteristics of the optical system would show as
deformations of the apparent positions of the reference stars. The
effect would be similar to studying the field of an astrograph with a
very large number of overlapping plates. Assuming a reasonable number

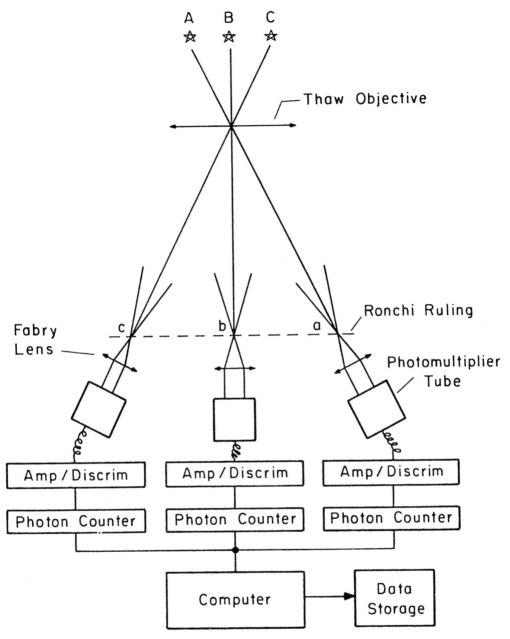

Figure 8 – The MAP will employ approximately a dozen light pickups, each transmitting light to an independent photomultiplier tube. The photon counters of each channel accumulate counts simultaneously, transmitting their counts to the central computer on the command of a single clock. The light pickups are clipped into masks made up for each survey region. Exact alignment is not important since the ruling and not the mask is the linear metric against which the star images are measured.

of imaging parameters, the precision with which the field could be mapped greatly exceeds that with which the measurements themselves are made. This adds directly to our ability to model the systematic errors of the optical system and presents the very real probability that existing astrometric telescopes will be able to achieve precisions beyond all expectations of their designers.

IMAGE TRAILER - A SINGLE-CHANNEL MAP

The Image Trailer (IT) was devised to confirm our approach to the concept of photoelectric centering and its implementation. This single-channel MAP records the modulation of the apparent intensity of a star image as it is trailed (by Earth's rotation) across a focal plane Ronchi ruling. The opaque lines, and the transparent lines between them, have widths approximately equal to the diameter of the star image, so the brightness variations plotted against time mimic a sine wave. Of course, the brightness variations are a function of the image's position on the ruling and thus relate time to focal plane position, or through the focal length of the telescope, time is related to angle. This is the principle of the MAP. Because it carries the most information, we define the center of the brightness wave as that point in time, approximately halfway between two minima, that divides the photon counts into halves. Hence we must observe the brightness as a function of time.

The IT accomplishes this with an equipment train consisting of a small 4-line pair per millimeter Ronchi ruling, an EMI bialkali (blue-sensitive photomultiplier tube, a Princeton Applied Research 1120 amplifier/discriminator, a PAR-1109 photon counter, an LSI-11 computer, an LA 36 Decwriter, and a paper tape reader/punch (fig. 9). The photomultiplier tube is encased in an uncooled housing with a Fabry lens positioned to focus a 37-mm^2 image of the telescope's objective on the 80-mm^2 cathode of the tube. (The dark current of this tube, uncooled, is only 10 counts/sec.) Between the Fabry lens and the focal plane of the telescope, within the tube housing, is a mirror and a pair of transfer lenses. These can be inserted into the beam to bring an image of the field to a reticled eyepiece (not shown in fig. 9).

To operate the IT, the telescope is pointed slightly to the west of a star and the electronics are activated. Next the refractor's main drive is turned off and the telescope is allowed to come to a rest position. The image of the star (approximately 0.12 mm in diameter in 1.5 arcsec seeing) now drifts across the focal plane at the rate

$$\frac{dx}{dt} = \frac{15}{14.60} \cos \delta \qquad (10)$$

where dx/dt is in millimeters per sidereal seconds, 14.60 is the mean focal plane scale of the 30-inch refractor in arcsec/mm, and δ is the apparent declination of the star. As the star image drifts onto the center of the reticle, the x guiding motor on the tailpiece (whose translator rate has been set for the declination of the star) is turned

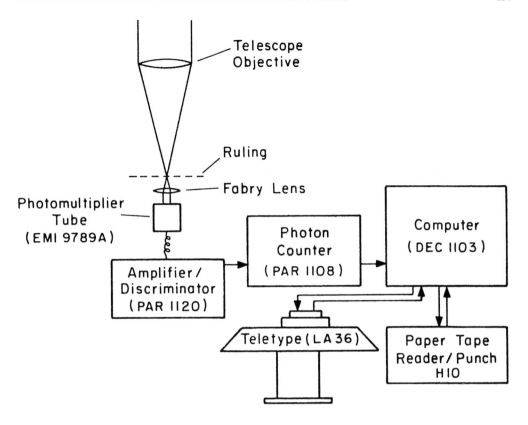

Figure 9.- Schematic of the components of the Image Trailer System

on. This motion will keep the image of the telescope objective near
the cathode's central area.

A star with a declination of approximately 13° drifts across 1 mm
of ruling per second of time, producing a 4-Hz signal. With the integration period of the photon counter set at 15 msec, approximately 17
samples of the apparent intensity are obtained for each pulse determination. For a 12.5 magnitude star the count rate varies from 1 or 2
counts/sample to nearly 60 as the star's image passes over the center
of a transparent line.

We define the pulse center, the time at which the image is central
to the transparent line, as

$$t_c = \frac{\Sigma \; c_i \; t_i}{\Sigma \; c_i}$$
(11)

where the summations are between successive minima, c_i is the count of
the ith sample, and t_i is the central time of the ith sample. Because
there are 4 line pairs per millimeter of ruling, the pulse centers
occur every 3.65 arcsec of image motion.

The ruling that modulates the light is a metal deposit on glass
with relative line spacings and edges accurate to better than 1 μm.
The IT's clock is based on a quartz crystal and the main motion of the
image is caused by Earth's rotation. The imaging characteristics of
the photographic refractor are well known and may be taken as strictly
gnomonic near the center of the field where the IT operates. Photon
statistics and the noise caused by dark counts do not become significant at magnitudes brighter than 16.7. Thus pulse errors larger than
a few microns or, approximately 30 mas, should be primarily due to the
atmosphere.

To determine the error of a pulse, the pulse centers (times of
central passage of the image over a transparent line) were fitted to a
linear model in time. This proved of sufficient precision, at all
observed declinations, for the short trails observed. The standard
deviations of the residuals to these least-squares solutions are plotted in figure 10 against pulse duration (effectively, the declination
of the region observed).

The success of the IT may be surmised from a comparison of two
features in figure 10 with characteristics of single star trails noted
earlier. First, the best fit line in figure 10 passes through 1 sec at
170 mas, insignificantly more precise than the average 1-sec standard
deviation of all the single star trail data. Second, the power law is
identical, the fourth root of t, to that of the single star trail plate
data. This agreement indicates that, like photographic plates, at these
short integration periods the errors caused by the detector are considerably less important than those caused by atmospheric seeing.

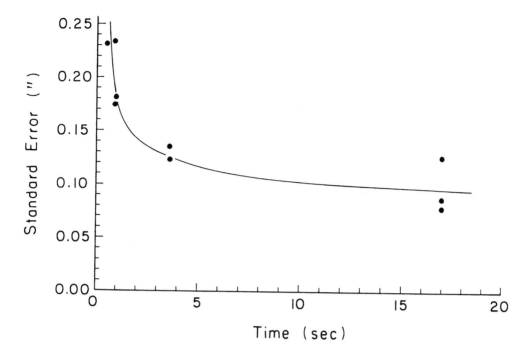

Figure 10 - The standard error of pulse centers about their best straight line fit indicates that data acquired with the image trailer is similar to data obtained from trail plates. The standard error of a 1-sec pulse is found to be 0.17 arcsec and the standard error decreases with the inverse of the fourth root of the integration period.

One area in which our tests indicate a significant advantage over the trail plate is in the magnitude of the trailed objects. A star fainter than 5th magnitude leaves a photographic trail too faint to measure. The IT, on the other hand, can track objects more than 10 magnitudes fainter. This gain is reasonable but still startling. Its implications are that a MAP-equipped Thaw refractor will be able to study 15th and 16th magnitude objects with virtually the same precision as it can measure objects many magnitudes brighter. This is certainly not the case in the current photographic program, where the number of exposures obtainable per night is highly dependent on the magnitude.

The success of the IT has shown that: (1) the relative focal plane position of a star image can be found photoelectrically, (2) star positions can be determined during an interval short in comparison to the time required for atmospheric seeing features to change, and (3) the high quantum efficiency of the photoelectric photometer will allow a MAP to measure, with a given telescope and at full precision, the positions of stars many times fainter than normally included on their photographic programs. For further discussion of the IT, see Stein (1978).

PRELIMINARY TESTS OF THE MAP

During August of 1979 a simplified 4 channel version of the MAP was placed on the Thaw Refractor. Using three channels to establish a reference frame we determined the relative position of a fourth star during a series of runs of varying duration. The purpose of these tests was to determine the MAP's precision as a function of integration time. The axis of measurement was approximately parallel to the celestial equator.

Figure 11 illustrates the results of this experiment. Plotted are the standard deviations, as determined from intracomparison of positions found from observations of ten different durations. We note that these seem to indicate that the error of an observation does decrease with the square root of the integration period. This is the behavior expected for data points free of short term systematic trends.

If the trend shown in Figure 11 holds for integrations of one-half hour, the error will be less than 3 mas. Thus a MAP-Thaw combination could determine nightly positions whose precision would exceed those now obtained by averaging the data acquired over an entire year of the photographic program. Such a system could also be used to determine stellar parallaxes with a precision exceeding 1 mas, making it possible to measure the trigonometric parallaxes of several galactic clusters and a number of supergiant stars directly.

Further tests are planned in the near future.

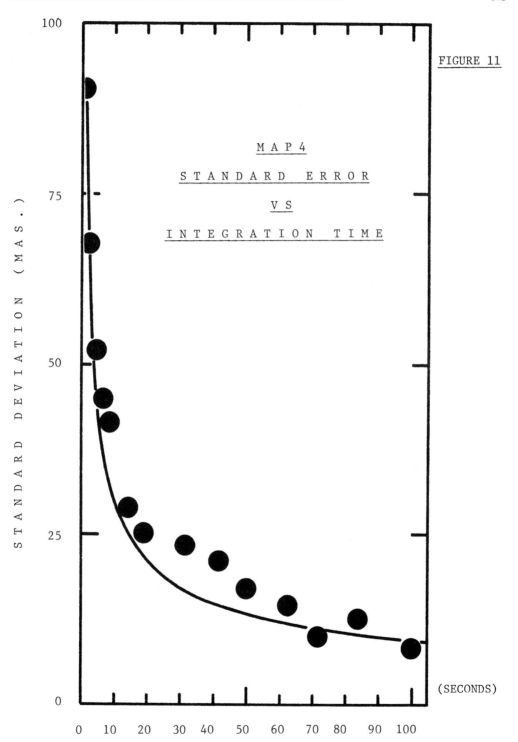

FIGURE 11

M A P 4

S T A N D A R D   E R R O R

V S

I N T E G R A T I O N   T I M E

COMMENTS ON ASTROMETRIC INTERFEROMETERS

With their ability to span many cells of atmospheric turbulence, long baseline interferometers offer some degree of immunity to the effects of astronomical seeing. Unfortunately, they are not readily adapted to high-accuracy astrometry, which requires the establishment of a suitable reference frame. Furthermore, the dimension that defines their advantage over seeing limits their application in the study of the nearest and most interesting stars.

The development of the interferometer from its original applications (Michelson, 1920) to a device of interest to astrometry began with a series of papers by Miller (1966, 1968, 1970, and 1971) with later modifications by Currie et al., (1974) and by Shao (1978). However, to our knowledge, the only interferometer designed specifically as an astrometric instrument resulted from the NASA-sponsored Stanford-Ames study on Project Orion (Black, 1980). Operating as an expansion of the astrometric imaging interferometer proposed by Gatewood (1976), the Orion device uses focal plane separations instead of the delay lines proposed by Miller. Although relatively massive and complex, this system offers several critical advantages over earlier proposals. Through its ability to determine focal plane positions, the 50-m interferometer is able to track the simultaneous positions of 20 or more stars in a given region. Thus it is able to model, much as is done in current long-focus astrometry, the effects of refraction and its anomalies directly, at the full precision of the observations. The detrimental effects of the optical system can be modeled from the apparent distortion it causes in the relative positions of these stars as they sweep through the field. Furthermore, the 20 or so reference stars define a catalog system in which the motion of the target star and each of the reference stars can be accurately defined.

In comparison, previously proposed systems would observe only one star at a time, either evaluating the direct and differential effects of refraction and its anomalies by the rapid successive observation of a handful of stars, or by observing the target object and a few reference stars in several widely spaced colors. The former technique has lost favor because of the need to observe at least seven stars every few minutes. This is a difficult procedure amidst the complexities of not only finding fringes but estimating their distance from the central fringe or, in one case, where one is on the fringe envelopes. The approach of observing in several wavelengths is more workable than that of shifting rapidly between several stars, but it carries with it a heavy penalty. The refractive indices of the observable colors only differ by approximately 1/50 of their mean value. Thus, to approximate the nonrefracted, above atmosphere position, the computer must extrapolate to a point that lies nearly 50 times outside the range of observation. Hence both the accidental and systematic errors of the observation are magnified by this factor. While some of the resultant increased random error may be overcome by repeated observation, no such reprieve is possible for the loss of systematic precision. Finally, these proposed interferometers are limited by time to a few reference stars in each region. Thus, the individual peculiarities of each star's motion are more difficult to find and model against their mean system. This presents the very real possibility of misinterpretation of the data.

Despite the advantages of the Orion concept, the size of the structure and the complexity of some of its subsystems suggest that lead times of many years would be necessary for successful funding and construction. Furthermore, interferometers suffer from a paradox. Generally, both their precision and the distance of source resolution vary approximately as the length of their baseline. (The situation is actually worse for baselines less than 50 m where the resolution increases much more rapidly with baseline than does astrometric precision.) Resolution is noted by the loss of fringe detection, with the signal being significantly reduced well before resolution (Anderson, 1920). Thus as one increases the baseline to gain precision he also increases the distance to the nearest stars for which the interferometer can determine astrometric positions. But equation (6a) implies, all other things being equal, that

$$m_d \propto \pi^{-1} \qquad (12)$$

Hence, the detectable mass is not generally a function of the baseline of the interferometer. As a consequence, if their practical problems can be overcome, interferometers will probably be employed in surveys to detect massive planets at relatively large distances, but will be of little use in the type of survey we propose below.

Equations (9) and (12) suggest that the smallest detectable masses will be found by diffraction-limited full apertures whose effective baseline is less than 10 m.

81-cm REFRACTOR

The instrument recommended in this section is a state-of-the-art variant of the 76-cm Thaw telescope. It is relatively inexpensive, approximately 10% of the cost of the Orion interferometer mentioned above, with a nightly precision per object approaching that projected for the interferometer and, because it relies on proven technology, it has a high probability of success.

Our approach has been to examine existent astrometric systems to determine their strengths and weaknesses and to find where six decades of technological evolution could improve their general design. The continuity gained in this manner lessens the chance of catastrophic oversights, while our possession of the arts of modern electronic and material engineering virtually assures substantial progress. The long success of existing systems can be attributed to several factors:

(1) Use of a single optical system for the simultaneous formation of the images of the target object and a moderate number of reference stars;

(2) Placement of all optical components at the pupil of the system;

(3) Simultaneous focal plane detection of the apparent relative positions and brightnesses of the target object and reference stars;

(4) Reduction of the observations using conditional equations with sufficient flexibility to model the significant effects of precision, nutation, stellar aberration, variations of latitude, general refraction, differential refraction, atmospheric scaling and/or displacement, variation of focal length, tilt of the optical axis and/or components, distortion, field curvature, and magnitude and/or color displacement and/or magnification.

Limiting these systems are observational errors arising from random variations in the detector and the atmosphere and the subjugation of the optical components to environmental variations. These errors enter at two levels. The first, and largest, are purely random in nature and can be diminished by repeated observation. The second error can affect groups of observations, slightly biasing their mean. The time extent of this second grouping ranges from those identified as "night errors" (Land, 1944) to the "discontinuities" associated with the occasional realignment of the optical components of a system (Gatewood and Eichhorn, 1973; Hershey, 1973; Gatewood, 1976; Heintz, 1976).

Figure 12 shows the major features of an optical system that shields the active components from the influences of its environment while incorporating the advantageous features of existing systems. To eliminate temperature variations in these components, they have been placed in a thermally controlled cell with an infrared coating on the front window of the system. Since a temperature difference between the glass and adjacent air would cause a deformation of the lens surface and cause local turbulence, the elements have been sealed in a light vacuum. To eliminate variations in the gravity vector, they have been placed so that their optical axis is parallel with Earth's axis of rotation and light from the target and reference stars are brought to the system by a single flat mirror. The necessary diameter of this mirror and its mounting requirements are minimized by placing it below the optical system. Made of a nearly temperature inert substance, the siderostat has a planned limited motion of $\pm 8^{\circ}$ from the meridian and only requires a declination motion of $\pm 18^{\circ}$ to survey 95% of a hemisphere. Like existing instruments, this one would (1) utilize a single optical system to image the simultaneous relative positions of the target and reference stars, (2) place all active optical elements at the system's pupil, (3) use a detector, the MAP, that simultaneously records the relative positions of the target and reference stars, and (4) allow the use of a truncated conditional equation to model all effects that can significantly affect the measured positions.

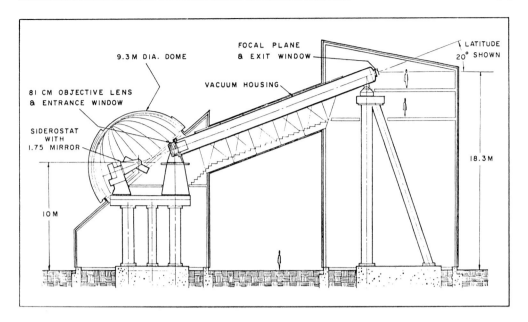

Figure 12 - The optics of the 81-cm refractor lie parallel to Earth's axis of rotation so that light can be brought into the system with only one flat mirror. This mirror is made smaller by placing it at the bottom of the optical axis. A refracting system is chosen so that the optical pupil is near this mirror as well as all other optical components. The secondary spectrum is kept within the Airy disk by a combination of glass choices and a limited bandpass (600 Å centered on 5000 Å). By fixing all nonzero power optical elements and placing them into a temperature-controlled vacuum, we have virtually removed all potential sources of systematic error. The figure of the flat, which has a limited range of motion and is made of temperature inert ceramic, can be held flat, controlled to approximately 1/20 of a wavelength.

Combined with the high quantum efficiency of the MAP, the 81-cm objective would gather sufficient photons for high-speed positional determination of stars as faint as visual magnitude 17, thus eliminating the need for a large field and the associated complex objective design. The field scale is 10 arcsec mm^{-1}, giving the system a focal ratio of 25.3. At such ratios all optical surfaces can be spherical and the curvatures slight. Secondary spectra could be kept within the Airy disk of the system over a bandpass of 4700 to 5300 Å by the proper choice of optical glasses, such as Schott SK11 and KZF2.

The siderostat is the only component more than a few percent of the focal length of the system from the optical pupil. Hence, variations in its surface do not affect all stars equally, but cause an apparent scale variation. The smoothness of this variation depends on the linear abruptness and magnitude of the surface features. Preliminary calculations indicate that abrupt variations of 0.05 wave can be modeled to better than 0.04 mas and, if constant, may be prereduced from the data. A feasibility study of this system conducted by the Boller and Chivens Division of the Perkin-Elmer Corporation indicated that such tolerances could be met. They noted that the 1.75-m circular flat (only an elliptical portion will be used) was actually easier to mount than a Cassegrain primary. An air flotation system with push-pull gravity-actuated side supports was recommended.

A SPACEBORNE ASTROMETRIC INSTRUMENT

Four sources of observational error are encountered by a spaceborne astrometric instrument, but we believe that, with careful attention to detail, their combined effect can be reduced below the signal produced by an Earth-mass planet orbiting within the ecoshell of a star at 12 parsec. The errors originate in the instrumentation, photon statistics, reference frame, and astrophysical phenomena that shift the photocenter of the primary star.

The latter, star spots for example, would cause small but perceptible shifts in the photocenter of the star. The effects of large spots can be modeled and reduced from the observations. These are distinct in that they have similar, probably short, periods and show for only half a rotation. Analysis of their motion will reveal the stellar rotation rate, the direction of the star's axis of rotation, and the latitude of the surface feature. Large numbers of small spots would cause a quasi-random variation in the apparent position of the photocenter which, in the case of the Sun, might mask the existence of the planet Mercury.

Another limitation is the stability of the reference frame. With the precisions sought, each reference star's motion will reveal a number of parameters. Depending on the complexity of its relative motion, it can be represented by focal plane equations similar to equation (4), to which terms will be added to model the orbital motion of the primary of an unresolved binary system. For extended observation at the potential precision of a spaceborne instrument, most of the reference stars will require all of these terms to describe their past, and thus to predict their future, positions. But the variance of the parameters is a function of the number of observations and will, if a sufficient number of reference stars are observed, gradually decrease so that eventually almost any desired precision can be obtained.

Photon statistics are the ultimate limitation on any observational efforts in that the positional data is carried by the electromagnetic radiation. Equation (9) shows that, all other things being equal, the theoretical precision of a spaceborne instrument goes as the second power of its aperture and approximately as the square root of its effective bandpass. Thus aperture and bandpass will be very important aspects of any spaceborne astrometric instrumentation.

Finally, there are limitations that depend on the design and engineering of the instrument itself. At first glance, these would seem to be the bane of the entire effort. However, if we return to the basic principles of high-precision astrometry outlined earlier and carry them to the extreme, we find the question of not whether such instruments can be built but rather when they will be built. The provisions for simultaneous observation and careful modeling will be similar to those on the ground and one optical system will image the entire region. However, we must now strictly enforce the condition that all optical components are at the pupil of the system. That portion of the optics that can be seen from all extremes of the instrument's field affects all stars similarly and, if the optical system is better than diffraction-limited, variations in the characteristics of the nonvignetted optics, unless they exceed approximately 1/8 wave, will not introduce new forms of systematic error. Thus there is a range in which a good optical surface may vary, causing only a variation in the magnitude of the terms, but not the form of the equation that models the imaging of the instrument.

This latter requirement determines the design of the instrument. Only a simple reflector has all of its optics directly in the pupil, and to be diffraction-limited over an extended field it must be of long focal ratio. The necessary focal length increases rapidly with aperture while the potential precision is a function of the aperture - hence the greater the required precision, the larger the required instrument.

Specifically, the optical component of the Spaceborne Astrometric Survey Instrument (SASI) would be formed from a single parabolized disk of zero expansion material such as CER-VIT with a diameter of approximately 2.1 m. To achieve a reasonable field in which the effects of coma and astigmatism are confined to the diffraction pattern of the aperture will require a focal ratio of approximately 20, or a focal length of 42m. Fortunately, the required satellite platform is quite light so that the several sections may be telescoped into each other and opened automatically after being placed in orbit by the Space Shuttle. The measuring device, similar in every respect to the MAP described earlier, is placed directly in the primary focal plane facing the aperture.

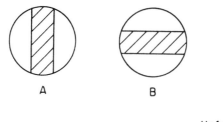

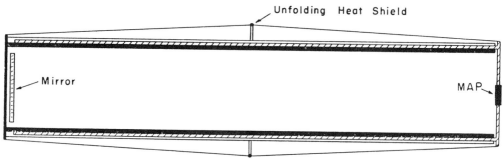

Figure 13 – The 2.1-m space-borne astrometric survey instrument
(SASI) requires an f/ratio of approximately 20 to bring its comatic and
astigmatic aberrations within the diffraction pattern of its aperture.
The MAP ruling will shadow nearly 1/3 of the mirror. To keep the pupil
on the mirror, an area slightly larger than that shadowed is covered by
a thin film on the mirror surface, A. This cover film could be recon-
figured to measure different coordinates without rotating the satellite,
B. The heat-shielded frame telescopes open to place the MAP-type detector in the focal plane after the satellite is launched from the Space
Shuttle. Support modules could be placed between the heat shield and
the instrument tube. The instrument carries the principles of high-
precision astrometry to the extreme, placing all optical components
exactly in the pupil where variations of less than 1/8 wave do not cause
new forms of astrometric error. The MAP-type detector is self-improving
in that its errors are easily determined from the measurements themselves
and can be removed. If utilized in a survey including the 81-cm refractors, the instrument may have the potential precision necessary to
detect every Earth-mass planet within the ecoshell of any solar-type
star within 12 parsec of Earth.

Because the ruling would otherwise define the inner edge of the aperture, the central portion of the primary under the ruling is covered along the axis being measured (fig. 12). The ruling spacing is approximately 140 μm. The bandpass will be 4000 Å, centered near 5000 Å, with a light-collecting area equal to that of a 1.37-m mirror, allowing the study of objects as faint as 20th magnitude. The perpendicularity of the ruling to the optical axis is less critical than it would be if it were not reduced by the equation of condition, but accurate focus is important. This may be achieved with a light frame of low expansion metal and near zero expansion ceramic spacers. Alternately, laser ranging active controls could maintain the focus with precision orders of magnitude greater than necessary.

The MAP seems to be the ideal detector for this spaceborne instrument. Its digitally encoded output is easily transmitted to the ground, and backups for its components are simplified by its redundancy. But, most important, the precision of the MAP can be matched to that of the photon statistics. As we have noted, the MAP is a self-improving measuring device. Every measuring session reveals as much about its metric, the Ronchi ruling, as about the star field studied. Nevertheless, there will be some error associated with our knowledge of the exact position of the ruling's lines. There is also an error associated with photon statistics. This is approximately the same as indicated by equation (9), from which we find, for a 15th magnitude star, there will be a measuring error of 0.3 mas/sec. If the central position of each ruling line is known to 1μm, the error caused by the ruling will equal that from photon statistics when we measure the position against 3 lines/sec. With the double forward and reverse motions of the ruling mentioned above, approximately 3 hr of measurements can be made with the ruling before the same lines are once again situated over the same field stars. During this period of independent measurements, the data stream will have a precision of 0.002 mas (2 μarcsec), approximately the size of the effect an Earth-mass planet in a 1-yr orbit would have on Alpha Centauri.

With its ability to make nearly continuous observations in orbit, the SASI may allow us to make a systematic search for Earth-like planets. From our admittedly biased point of view, to be Earth-like implies to be of similar mass and temperature. (Note that we have defined Jupiter-like as having mass and period similar to Jupiter's.) This requires us to include an estimate of the primary star's illumination in our calculations, but it gives us the ability to estimate the difficulty or ease of detecting a planet, if orbiting a given star, on which life might thrive. Of most interest are the main sequence stars. For these, assuming the mass-luminosity relationship of Harris et al. (1963), we find that for stars whose absolute magnitude is less than 7.5, the size of the perturbation induced in the apparent motion of the primary star is

$$A = 3 \times 10^{-6} \, \pi \, M \quad \text{(arcsec)} \qquad (13a)$$

from which we note that Jupiter is 1650 times easier to detect than
Earth. Similarly, we may write a detection index for Earth-like planets
as

$$I_\oplus = \pi M \tag{14a}$$

The period P of such a planet's orbit is approximately $P = M^{3.5}$, where
M is in solar masses and P is years.

For K and M stars, whose absolute magnitude exceeds 7.5, we have

$$A = (2.3 \times 10^{-6}) \pi M^{0.4} \tag{13b}$$

and

$$I_o = 0.76 \ \pi \ M^{0.4} \tag{14b}$$

where Earth-like planets have orbital periods of $P = 0.66 \ M^{2.6}$ yr. It
is interesting to note that these equations show that Earth-like planets
are easier to detect orbiting more massive, main-sequence stars.

The principles employed in the SASI are so straightforward it is
hard to see how they might fail. Even if the system does not achieve
the projected precision, it should exceed values projected for other
proposals by one or two orders of magnitude.

CONCLUSIONS

Since the presence of planetary systems can be detected astrometrically from any orientation of the observer and orbital plane, a
meaningful study of their frequency and basic characteristics is
possible. A survey of the nearby stars would reveal the case-by-case
characteristics of their planetary systems and most of their individual
planets. Negative results would yield the range of masses and orbital
periods not found, as well as statistical confidence levels. As a group,
both the positive and negative results will give us the statistics
necessary to better understand planetary systems in general and thus
our own in particular. But perhaps more importantly, it will give us
our first detailed picture of our vicinity of the galaxy.

The current survey, which now employs the active processing of
systematic errors to the precision of the normal points, has sufficient
accuracy to yield detectable masses a fraction of that of Jupiter
orbiting any of 20 nearby stars. Tests show that the basic concepts
behind the MAP are sound, and we estimate that its implementation will
increase our precision by five times and allow the search for Jupiter-
like planets orbiting any of nearly 100 different stars. However, to
obtain a statistically significant sample, including primary stars of
most spectral types, further new instrumentation will be necessary.
The current survey relies on a single optical system that is fully
exposed to all variations of its environment. Thus the kind and magnitude of the processes that affect image position can vary over relatively

short periods of time. This in turn limits the unified processing of
these parameters to relatively small groups of data and hence restricts
the precision of the final normal point. Both the 81-cm refractor and
the 2.1-m simple reflector are shielded from the changes that affect
current instruments and promise precisions of 10 times and more than
10,000 times higher, respectively.

Equation (12) indicates that the best sample is the nearest, while
the probable program size of a pair of 81-cm refractors (one in each
hemisphere) is several hundred stars. This suggests a survey of the
stars within 12 Parsec of the Sun. With the exception of Population
II stars, which are probably deficient in terrestrial planets and organic
compounds, this volume of stars is homogeneous. The 81-cm, 12-parsec
Survey Instruments (12ψ) would prepare the way for the 2.1-m spaceborne instrument. With a seasonal precision near 0.2 mas, they could
detect virtually any Jupiter-like planet in the sample and, in combination with a radial velocity study, detect almost all of the unresolved
binary stars within this volume. This effort would give us our first
estimates of the true frequency of planetary systems and of the number
of stars not members of multiple-star systems.

With this preparation, the spaceborne instrument would have a
somewhat smaller program than the 81-cm systems and could concentrate
its efforts toward discovering Earth-like planets. Studies by Harrington
and Harrington (1978) indicate that these Earth-like planets can occur
in a wide variety of, but not all, stellar systems. Thus the efforts
of the 12ψ are very meaningful, reducing both the time span and costs
of the space mission. In somewhat the same fashion, the current survey
and the MAP-assisted survey would reduce the effort of the 12Ψ phase,
troubleshooting both design and data processing problems.

The current lack of known extrasolar planetary systems is
probably more indicative of the care necessary in our quest to find
them than of their absence. However, the precisions now being achieved
and those indicated for the systems now being designed and tested
promise a better knowledge of the frequency of the systems thought
necessary to sustain life as we know it.

ACKNOWLEDGEMENTS

Without the widespread encouragement and assistance of many
persons, this effort would not have been possible. In particular we
appreciate the encouragement of Drs. Morris Aizenman, John Billingham,
David Black, Myron Garfunkel, Jesse Greenstein, David Morrison, Philip
Morrison and Carl Sagan. Among many others we benefited from discussions
with: Drs. James Baker, Ronald Bracewell, Clyde Chivens, Douglas Currie,
Frank Drake, Heinrich Eichhorn, Robert Harrington, Wulff Heintz, Roger
Howe, Charles KenKnight, Larry Lesyna, Richard Miller, Krzysztof
Serkowski, Mike Shao, Orestes Stavroudis, Kaj Strand, Arthur Upgren,
John Vorreiter, and Chester Wheeler. Financial support has been

received from The National Aeronautics and Space Administration through grant number NSG 2117 and from the National Science Foundation through grant number AST 75-02641-A01.

REFERENCES

Abt, H.A. and Levy, S.G. (1976) Multiplicity among solar type stars. Astrophys. J. Suppl. 30, 273.

Anderson, J.A. (1920) Application of Michelson's interferometer method to the measurement of close double stars. Astrophys. J. 51, 263.

Black David C., editor (1980) Project Orion, Design Study of a System for Detecting Extrasolar Planets, NASA SP-436.

Connes, P. (1978) A proposed ground-based narrow field astrometric system. Coll. on European Satellite Astrometry (in print).

Currie, D., Knapp, S. and Liewer, K. (1974) Four stellar-diameter measurements by a new technique: amplitude interferometer. Astrophys. J. 187, 131.

Drake, F. (1975) Minutes of first planetary detection workshop (edited by J. L. Greenstein), Unpublished report.

Eichhorn, H. and Williams, C. (1963) On the systematic accuracy of photographic astrometric data. Astron. J. 68,221.

Frederick, L. W., McAlister, H.A., van Altena, W.F. and Franz, O.G. (1974) Astrometric Experiments for the Large Space Telescope. ESRO Symposium on Space Astrometry, Frascati, Italy.

Gatewood, G. (1974) An astrometric study of Lalande 21185. Astron. J. 79,52.

Gatewood, G. (1975) Minutes of the first planetary detection workshop (edited by J. L. Greenstein), Unpublished report.

Gatewood, G. (1976) On the astrometric detection of neighboring planetary systems. Icarus 27,1.

Gatewood, G., and Eichhorn, H. (1973) An unsuccessful search for a planetary companion of Barnard's Star (BD + 4° 3561). Astron. J. 78, 777.

Gatewood, G. and Russell, J. (1974) Astrometric determination of the gravitational redshift of van Maanen 2 (EG5). Astron. J. 79, 815.

Gatewood, G., and Russell, J. (1979) Astrometric studies of ε Eridani and τ Ceti (to be published)

Gould, B.A. (1866) On the reduction of photographic observations. Mem. Nat. Acad. Sci. IV. 173.

Harrington, R.S. and Harrington, B.J. (1978) Can we find a place to live near a multiple star? Mercury 7 , 34.

Harris, F., Strand, K. Aa. and Worley, C. (1963) Empirical data on stellar masses, luminosities, and radii. Basic Astronomical Data (edited by K. Aa. Strand), University of Chicago Press.

Heintz, W. (1976) Systematic trends in the motions of suspected stellar companions. M.N. 175, 533.

Heintz, W. (1978) Astrometric study of four binary stars. Astron. J. 77, 160.

Hershey, J. (1973) Astrometric analysis of the field of AC + 65° 6955 from plates taken with the Sproul 24-inch refractor. Astron. J. 78, 421.

Hudson, C. (1929) Irregularities in refraction. Publications of the Allegheny Observatory 6, 1.

Huang, Su-Shu (1973) Extrasolar planetary systems. Icarus 18, 339.

Kumar, S. (1963) The Helmholtz-Kelvin time scale for stars of very low mass. Astrophys. J. 137, 1126.

Kumar, S. (1972) Hidden mass in the solar neighborhood. Astrophys. Space Sci. 17, 219.

KenKnight, C., Gatewood, G., Kipp, S. and Black, D. (1977) Atmospheric turbulence and the apparent instantaneous diameter of the Sun. Astron. and Astrophys. 59, L27.

Land, G. (1944) Systematic errors in astrometric photographs. Astron. J. 51, 25.

Levinson, F. and Ianna, P. (1977) Emulsion shifts and the astrometric accuracy of the photographic plate. Astron. J. 82, 299.

Lippincott, S. (1957) Accuracy of positions and parallaxes determined with the Sproul 24-in. refractor. Astron. J. 62, 55.

Lippincott, S. (1978) Space Science Reviews, 22, 153.

Lippincott, S. and Hershey, J. (1972) Orbit, mass ratio, and parallax of the visual binary Ross 614. Astron. J. 77, 679.

Michelson, A. A. (1920) On the application of interference methods to astronomical measurements. Astrophys. J. 51, 257.

Miller, R. (1966) Measurements of stellar diameters. Science 153, 581.

Miller, R. (1968) A modest Michelson stellar interferometer. Astron. J. 73, 5108

Miller, R. (1970) A fringe detector for use with Michelson stellar interferometer. AURA Engineering Tech. Rep. 29.

Miller, R. (1971) A 100-meter Michelson stellar interferometer. AURA Engineering Tech. Rep. 40.

Russell, J. L. (1976) The Astrometric Standard Region in Praesepe (M44). Ph.D. Dissertation, University of Pittsburgh.

Schlesinger, F. (1916) Irregularities in atmospheric refraction. Publications of the Allegheny Observatory 3, 1.

Schlesinger, F. (1917) Photographic determinations of the parallaxes of fifty stars with the Thaw refractor. Publications of Allegheny Observatory 4, 1.

Shao, M. (1978) Optical interferometers in astrometry. I.A.U. Colloquium No. 48, Vienna, Austria.

Stein, J. W. (1978) Development of the image trailer, a prototype of the multichannel astrometric photometer. Ph.D. Dissertation, University of Pittsburgh.

Strand, K. Aa (1977) The triple system, Stein 2051 (G175-34). Astron. J. 82, 745.

van Altena, W. (1974) The Yerkes Observatory photoelectric parallax scanner. I.A.U. Symposium 61, 311.

van de Kamp, P. (1975) Unseen astrometric companions of stars. Ann. Rev. Astron. and Astrophys. 13, 295.

SEARCH FOR PLANETS BY SPECTROSCOPIC METHODS

K. Serkowski[1]
Lunar and Planetary Laboratory and Steward Observatory
University of Arizona, Tucson

The Sun moves around the barycenter of the Sun + Jupiter system
with an average velocity of 12.7 m/s. Therefore, to detect a Jupiter-
like planet around a solar-type star by spectroscopic methods, the
annual average of measured velocity should have a mean error not larger
than about +5m/s. This means that accuracy should be almost two orders
of magnitude better than commonly achieved for stellar radial velocities.

The highest precision of radial velocity is obtained in a given
observing time if the dispersion is sufficient to fully resolve the
spectrum (Campbell and Walker, 1980). For stars similar to the Sun
this happens when the spectral resolution element is about 0.07 Å wide.
If the photometric accuracy is limited by photon statistics, the
weight (i.e. the inverse square of mean error) with which a spectral
resolution element contributes to determination of radial velocity is
proportional to the signal I obtained from this spectral resolution
element and is proportional to square of the derivative $dI/d\lambda$ of this
signal over wavelength (Serkowski, 1978). If each spectral resolution
element corresponds to a single picture element (pixel) of a detector
array on which the spectrum is imaged, the total weight of the radial
velocity measurement is proportional to the number of pixels used; this
remains true even if we place a Fourier spectrometer in front of the
spectrograph which images spectrum on the detector array. To increase
the number of pixels we should use an echelle spectrograph and a two-
dimensional detector.

Radial velocities can be also obtained with high accuracy using a
single detector (photomultiplier). A mask is placed along the spectrum
which transmits first only the portions of spectrum on the ascending
branches of stellar absorption lines, then only those on the descend-
ing branches (Connes, 1980, Serkowski et al., 1979b). However, in such
a correlation spectrometer only those portions of spectrum for which an
absolute value of the derivative $dI/d\lambda$ is very large can be used; if
the mask also transmits portions of the spectrum with smaller values of
this derivative they contribute more to noise than to the signal and
they degrade the precision of radial velocity. For that reason, the

M. D. Papagiannis (ed.), Strategies for the Search for Life in the Universe, 155–161.
Copyright © 1980 by D. Reidel Publishing Company.

correlation spectrometer utilizes only a fraction of radial velocity information contained in the stellar spectrum.

Let us consider again a spectrum imaged on a two-dimensional detector array with certain limited number of pixels. Precision of radial velocity will obviously be higher in the blue spectral region where the absorption lines in solar-type stars are crowded and an average value of the square of the derivative $dI/d\lambda$ is higher than in the red part of the spectrum. On the other hand, silicon detectors are more sensitive in the red portion of the spectrum than in the blue and later-type solar-type stars are much brighter in the red compared to the blue part of the spectrum. Most importantly, since in the red part of spectrum the stellar absorption lines are usually separated by clear patches of continuum, we may transmit the stellar light through an absorption cell filled with a gas producing absorption lines which will not be blended with stellar lines. Campbell and Walker (1980), who utilize this technique to search for extrasolar planets, found that hydrogen fluoride is the most suitable gas for that purpose. A detector should be very linear because, to eliminate any residual effects of blending, the stellar spectrum obtained without the absorption cell has to be subtracted from the spectrum taken through the absorption cell, used as a wavelength standard.

In the University of Arizona radial velocity spectrometer (Serkowski et al., 1979a, 1979b, 1980) a spectral region 250 Å wide around the wavelength 4250 Å is used, in which the amount of radial velocity information per unit wavelength is maximized for the solar-type stars. Because of crowding of stellar absorption lines in this spectral region an absorption cell cannot be placed in the stellar light beam. Instead, an absorption cell containing nitrogen dioxide is placed in a beam of light from an incandescent bulb. Obviously, accurate radial velocities can be obtained only if the stellar light beam and the light beam from the bulb are passing in exactly the same way through the spectrometer optics. In particular, they should illuminate the spectrometer entrance aperture in exactly the same way.

To achieve this, the two light beams are alternately focused on an input end of a single fused silica fiber, about 10 meters long (cf. Angel et al., 1977, Connes, 1980). A plastic-clad fiber of 0.125mm core diameter, manufactured by Galileo Electro-Optics Corp., is used, for which no other degradation of the focal ratio of incident beam is found than that caused by diffraction on the fiber aperture (Heacox, 1980). The fiber acts as an image scrambler: illumination of the output end of the fiber is almost uncorrelated with illumination of the input end, on which a star image is formed by telescope optics. The light losses in the fiber are small. The scrambling is most efficient if the input end of the fiber is at the prime focus of the telescope, a position which also minimizes the light losses in the telescope optics. The output end of the fiber enters the spectrometer which stands on the floor in the telescope dome. A small lens, cemented to the output end of the fiber, makes the focal ratio of the outgoing beam independent of that for the beam entering the fiber. This is particularly

important in conditions of poor seeing when the stellar light reflected
from the outer zones of the telescope mirror is more likely to be cut
off by the edges of the input end of the fiber than the light reflected
from the inner zones of the mirror, at imperfect telescope focusing.

Very high accuracy wavelength calibration is difficult to achieve
if the wavelength corresponding to any particular pixel of a solid
state detector array is defined by the position of the center of this
pixel. Non-uniform sensitivity over the pixel area will cause a shift
of the photometric center of a pixel relative to its geometric center.
This shift will have to be known to an accuracy of a few thousandths of
pixel size to get radial velocity accuracy needed for detecting extrasolar planets. Also, the spectrum should not shift by more than few
thousandths of a pixel between observations of a star and that of a
standard lamp with an absorption cell. This may be difficult or impossible to achieve.

To avoid having the wavelength defined by the position of the
spectrum on the detector we may insert a Fabry-Perot interferometer or
a Michelson interferometer (Fourier spectrometer) in front of an echelle
spectrograph. Wavelength will then be defined by the path difference
within the interferometer. The measurements will be made for about 20
values of this path difference. The interferometer will increase the
spectral resolution by a factor of about 10, allowing the use of a
smaller echelle grating and a smaller detector. If photometric accuracy is limited by photon statistics, both Fabry-Perot and Michelson
interferometers will be approximately equally efficient for our purpose
(Serkowski, 1978); if detector readout noise is a principal source of
error, the Michelson interferometer is preferred because of its multi-
plexing advantage. For high precision wavelength calibration the light
path of the calibrating laser beam in the Michelson interferometer
should preferably be exactly the same as the path of the stellar light
beam. If a Fabry-Perot interferometer is used, it should be operated
in a vacuum and should have spacers optically contacted to interferometer plates. Both the plates and the spacers should be made from a
material of negligibly small thermal expansion, for example from
Corning Titanium Silicate for which the thermal expansion coefficient
changes its sign at room temperature. The interferometer should be
kept in a thermostat at a temperature at which this change of sign
occurs.

These requirements are fulfilled for the Fabry-Perot etalon used
in the University of Arizona radial velocity spectrometer. The instrument has a transmission grating-prism (grism) placed in a collimated
light beam between the interferometer and the echelle spectrograph.
Therefore, about 7 echelle orders are imaged on the detector, each
order consisting of a series of points corresponding to individual
transmission maxima of the Fabry-Perot etalon. These transmission
maxima are separated by about 0.6 Å, which corresponds to about 5
pixels on a two-dimensional solid state detector array. The interferometer finesse is about 10. Therefore, to obtain a complete spectrum,

measurements are made at about 20 different tilts of the Fabry-Perot
etalon. At maximum tilt the wavelength of each Fabry-Perot transmission maximum is shorter by about 0.7 Å than at zero tilt. The tilts
are calibrated with a hollow-cathode lamp.

In the future we are planning to improve the transmittance of
this instrument by utilizing the light reflected from a Fabry-Perot
etalon, which is presently lost. This is explained in Figure 1, where
fiber 1 is the optical fiber connecting the telescope to the spectrometer. The light emerging from this fiber, after passing through a
collimator, a Fabry-Perot etalon, a grism, and a decollimator, forms
a short spectrum on slit 1 at the input of the echelle spectrograph.
The light from fiber 1 which is reflected from the etalon is focused
by the collimator on the input end of fiber 2. The light from the
output end of fiber 2 which is transmitted through the etalon forms a
spectrum on slit 2; that which is reflected from the etalon is focused
on the input end of fiber 3. The light from the output end of fiber 3
forms a spectrum on slit 3. The input ends of the fibers 2 and 3 are
moved by a stepper motor synchronously with a stepper motor operating
the tilt changes of the Fabry-Perot etalon; Figure 1 shows these input
ends at two positions corresponding to two tilts of the etalon. The
output ends of all three fibers have fixed positions. Because of different angles of incidence on the Fabry-Perot etalon, the wavelengths
of Fabry-Perot transmission maxima for the light passing through slit
2 are shorter by 0.2 Å than those for the light passing through slit 1.
The transmission maxima for the light from slit 3 have wavelengths
shorter by 0.4 Å than those for the light from slit 1. Therefore, the
range of etalon tilts can be smaller than in the present instrument
and we only need to make measurements at 7 tilts, instead of 20. The
detector array must be three times wider in the direction of cross-
dispersion. Going in this direction, first the 7 echelle orders from
slit 3 will be imaged, then the 7 orders from slit 1, and finally those
from slit 2. Taking into account the light losses, we should gain at
least a factor of 2 in the amount of stellar light utilized.

When the reflecting surfaces of the Fabry-Perot etalon are separated by optically contacted spacers, they are usually not exactly plane
parallel, but form a wedge amounting to a small fraction of an arc-
second. During scanning of an emission line from a hollow cathode lamp,
the area of the etalon transmitting most of the light moves on its surface when we change the etalon tilt. Therefore, if the spectrum is not
very sharply focused on the detector, an image of the emission line
will shift on the detector when the etalon tilt is changed. Since
similar shifts occur also for stellar absorption lines, they may cause
serious systematic errors in radial velocities which may be difficult
to eliminate. Such errors are caused not only by imperfect focusing of
the spectrum on the detector, but also by astigmatism in the echelle
spectrograph. An astigmatic circle of least confusion formed on the
detector is a mirror image of the etalon. Following a suggestion by
C. Sepulveda, we are eliminating any residual astigmatism by placing
a tilted plano-convex lens below the entrance slit to the echelle
spectrograph.

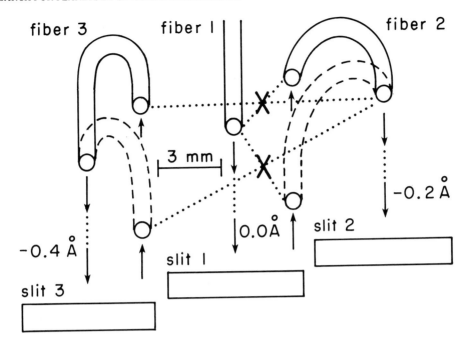

Fig. 1. Utilizing the light reflected from a Fabry-Perot etalon. All fiber ends are in one plane perpendicular to the axis of the instrument. The points of this plane which would be imaged on themselves by the light emanating from these points and reflected by the etalon (at two etalon tilts) are denoted by x. The positions of the fibers at two etalon tilts are indicated by solid and dashed lines. Arrows denote the directions of light. The fibers and arrows are shown in perspective: the fibers are supposed to be above the plane of the drawing, arrows are supposed to be perpendicular to the plane of the drawing. The slits are shown in a different plane than the fiber ends.

Below the same entrance slit we are also placing a sub-stepping wedge of fused silica which helps to eliminate the effects of rapid changes in signal, caused, e.g., by atmospheric seeing, and the effects of differences in pixel sensitivity. At each etalon tilt a star and a standard lamp are measured at two positions of this sub-stepping wedge, denoted by a and b. The tilt of the sub-stepping wedge is changed so that the signal $I(m,a)$ in interference order m at substep a is measured by the same group of detector pixels as the signal $I(m+1,b)$. The ratio

$$\frac{I(m,a)}{I(m-1,a)} \bigg/ \frac{I(m+1,b)}{I(m,b)} \qquad (1)$$

is independent of rapid time changes in the signal and independent of differences in the pixel sensitivity. For convenience we treat the quantity

$$S(m) = \ell n \left[\frac{I(m,a)}{I(m-1,a)} \cdot \frac{I(m,b)}{I(m+1,b)} \right]^{1/2} \tag{2}$$

as the signal in interference order m.

To eliminate the effects of small nonlinearity of the detector we replace I in eq. (2) by a quantity $(1+AI)I$, where I is observed signal and A is a constant much smaller than 1. Since $\ell n(1+AI) \simeq AI$, we have

$$S \cong \frac{1}{2} \ell n \left[\frac{I(m,a)}{I(m-1,a)} \frac{I(m,b)}{I(m+1,b)} \right] +$$

$$+ \frac{A}{2} [I(m,a) + I(m,b) - I(m-1,a) - I(m+1,b)], \tag{3}$$

where A, which is assumed to be the same for all transmission maxima, is found by comparing a given observation of a star with a mean normalized spectrum $\overline{S}$ for this star. For this mean spectrum we assume that by definition $A = 0$. To eliminate any residual effects of nonlinearity, we make for each stellar observation the sum of weights of all values of S for which $dS/d\lambda > 0$ equal to the sum of weights for $dS/d\lambda < 0$.

Similarly, contamination of each Fabry-Perot maximum by light from the neighboring transmission maxima can be eliminated by replacing each of the four signals $I(m)$ in eq. (2) by a quantity $I(m)-CI(m-1)-CI(m+1)$. Eq. (3) now becomes

$$S \simeq \frac{1}{2} \ell n \left[\frac{I(m,a)}{I(m-1,a)} \frac{I(m,b)}{I(m+1,b)} \right] +$$

$$+ \frac{A}{2} \left[I(m,a) + I(m,b) - I(m-1,a) - I(m+1,b) \right] +$$

$$+ \frac{C}{2} \left[\frac{I(m-2,a) + I(m,a)}{I(m-1,a)} + \frac{I(m,b) + I(m+2,b)}{I(m+1,b)} - \right. \tag{4}$$

$$\left. - \frac{I(m-1,a) + I(m+1,a)}{I(m,a)} - \frac{I(m-1,b) + I(m+1,b)}{I(m,b)} \right] .$$

Comparing the first terms in eq.(4) for a given stellar observation, calculated from this equation, with the values of $\overline{S}$ for the mean spectrum of this star for which constants A and C are assumed to be zero, we find the values of these two constants from a least squares solution.

This discussion suggests that measuring stellar radial velocities to accuracy sufficient for detecting extrasolar planets is within the present state of the art. We hope that for solar-type stars the intrinsic changes in radial velocity are negligibly small for periods longer than several hours. This hope is based on five years of observations of a solar potassium line at 7699Å with a resonant scattering solar spectrometer, which were made by J. R. Brookes et al. (1978) in Birmingham. As reported by G. Isaak to the Stellar Pulsation Workshop at the University of Arizona in March 1979, these data reveal no long-term (periods longer than 5 hours) solar radial velocity variations of amplitude 0.5m/s or greater. Unfortunately, because of the orbital motion of the Earth, this very accurate technique cannot be applied to measuring stellar radial velocities.

Our project is supported by grants from the National Geographic Society, National Aeronautics and Space Administration, and National Science Foundation.

REFERENCES

Angel, J.R.P., Adams, M. T., Boroson, T. A. and Moore, R. L.: 1977, Astrophys. J., 218, 776.
Brookes, J. R., Isaak, G. R., McLeod, C. P., van der Raay, H. B. and Roca Cortes, T.: 1978, Mon. Not. R. Astr. Soc., 184, 759.
Campbell, B. and Walker, G.A.H.: 1980, Publ. Astron. Soc. Pac. (in press).
Connes, P.: 1980, in D. C. Black and W. E. Brunk (ed.) "An Assessment of Ground-Based Techniques for Detecting Other Planetary Systems" NASA Conference Proceedings (in press).
Heacox, W. D.: 1980, in A. Hewitt (ed.) Proceedings of "Optical and Infrared Telescopes for the 1990's" (Tucson; in press).
Serkowski, K.: 1978, in M. Hack (ed.) "High Resolution Spectrometry" (Trieste) 245.
Serkowski, K., Frecker, J. E., Heacox, W. D., KenKnight, C. E. and Roland, E. H.: 1979a, Astrophys. J. 228, 630.
Serkowski, K., Frecker, J. E., Heacox, W. D. and Roland, E. H.: 1979b, in D. L. Crawford (ed.) "Instrumentation in Astronomy III" Proc. SPIE 172, 130.
Serkowski, K., Frecker, J. E., Heacox, W. D. and Roland, E. H. : 1980, in D. C. Black and W. E. Brunk (ed.) "An Assessment of Ground-Based Techniques for Detecting Other Planetary Systems" NASA Conference Proceedings (in press).

[1] Presented by Ian S. McLean.

THE SEARCH FOR PLANETS IN OTHER SOLAR SYSTEMS THROUGH USE OF THE SPACE TELESCOPE

William A. Baum
Lowell Observatory, Flagstaff, Arizona 86002, U.S.A.

ABSTRACT. The Space Telescope can make important contributions to the search for extrasolar planetary systems. This note particularly calls attention to the capability of the CCD camera system and discusses. criteria for the selection of candidate stars.

The search for planets around stars seems destined to be an exciting adventure with high potential for public interest and important implications for the future of space exploration. Three instruments aboard the Space Telescope have been mentioned as possibly contributing to this search: the CCD camera, the fine-guidance system, and the faint-object camera. The CCD camera, also known as the wide-field/planetary camera system, is expected to provide high-quality imaging to 28th magnitude and may achieve astrometric precision to 1 or 2 milliarcseconds for stars down to 22nd magnitude. The fine-guidance system does not provide imaging but operates astrometrically down to about 17th magnitude with a design goal of 2 milliarcseconds. The faint-object camera is expected to provide high-quality imaging to the same magnitude limit as the CCD camera but has a rather small field for finding enough reference stars for astrometry.

Under ideal conditions, the Space Telescope cameras should theoretically detect <u>images</u> of extrasolar planets directly in a few special cases, but I am personally <u>not</u> optimistic about direct image detection. It demands holding scattered light down to an unusually low level.

On the assumption of long-term stability in CCD pixel geometry, the most promising Space Telescope method seems instead to be the astrometric measurement of candidate stars on CCD-camera images obtained from time to time during the first few years of Space Telescope operation. Such images should, of course, be first scrutinized for any direct evidence of faint companions (not necessarily substellar), but an unambiguous answer concerning the existence or non-existence of other planetary systems like our own will probably have to come from astrometric measurement of those images.

M. D. Papagiannis (ed.), Strategies for the Search for Life in the Universe, 163–166.
Copyright © 1980 by D. Reidel Publishing Company.

Precision astrometry to 22nd magnitude means that there will be reference stars within a small angular distance of each program star. But more important, the irregular motions of <u>faint</u> reference stars (due to their companions) will typically be smaller than the irregular motions of the program star that we seek to measure. In the jargon of astrometry, the "cosmic errors" will be acceptably small.

Figure 1 shows the expected precision for the centroiding of star images obtained with the Space Telescope CCD cameras during exposure times of 1000 seconds. The CCD camera system includes a choice of two different image scales, 100 milliarcseconds per pixel at f/12.9 or 43 milliarcseconds per pixel at f/30. Although f/30 provides higher precision per star, the f/12.9 scale provides a field five times larger in sky area, so more reference stars would be available. Bars at the left-hand ends of the curves in Figure 1 indicate approximate saturation magnitudes for 1000-second exposures. For shorter exposures, these curves (and the saturation magnitudes) shift toward the left.

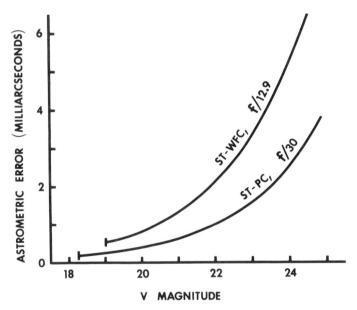

Figure 1. Precision with which the centroids of star images should be determinable from 1000-second exposures with the Space Telescope CCD cameras.

As a criterion of astrometric detectability of planetary systems, I have imagined each nearby star to possess a hypothetical "Jupiter" (a planet of Jupiter's mass and orbital size) and have calculated the resulting amplitude of positional oscillation of the star that would be expected. Figure 2 is a plot derived from W. Gliese's 1969 "Catalog of Nearby Stars," kindly made available in machine-readable form by J. A. Westphal. Absolute magnitudes are based on a mean mass-luminosity relation, satisfactory for this statistical exercise. The family of upward

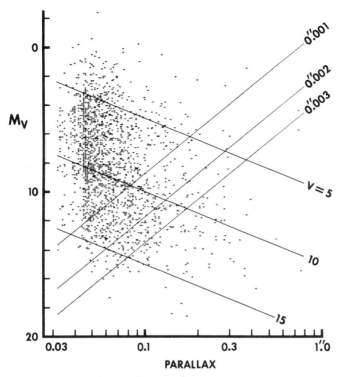

Figure 2. The estimated absolute magnitude distribution of nearby stars in W. Gliese's 1969 catalog as a function of parallax. The sloping lines represent limits for the detection of planetary systems, as explained in the text.

sloping lines represents half-amplitudes of positional oscillation that the hypothetical "Jupiters" would produce at those loci in Figure 2. One may think of them as <u>upper</u> limits for the selection of planetary-system search candidates. The downward sloping lines represent loci of stars having the indicated values of apparent magnitude V and are therefore <u>lower</u> limits for detection. For any particular choice of upper and lower limits, the stars falling within the wedge-shaped area at the right of those limits are the planetary-system candidates of interest.

Figure 3 then shows how the number of "Jupiter" detection candidates will depend on the astrometric accuracy and magnitude threshold of the instrument. If the CCD cameras succeed in reaching a long-term accuracy of 1 milliarcsecond, we could evidently choose an observing list from among more than 500 candidate stars. For a more conservative selection, there are about 100 candidates at the 3-milliarcsecond level. In early Space Telescope imaging, that list might wisely be reduced to about ten prime cases having particularly favorable distributions of reference stars. Among those prime cases, we should then be able to determine with a high degree of certainty whether or not planetary systems even vaguely similar to our own are present.

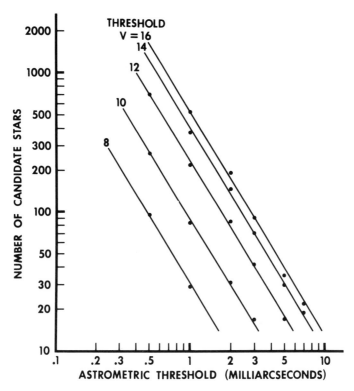

Figure 3. The number of stars for which the astrometric detection of Jupiter-like planets should be possible, as a function of astrometric accuracy and magnitude threshold.

A COMPARISON OF ALTERNATIVE METHODS FOR DETECTING OTHER PLANETARY SYSTEMS

David C. Black
Space Science Division
NASA Ames Research Center
Moffett Field, CA 94035

Detection of other planetary systems is a demanding observational task that will require significant improvements in instrumentation. There are two general techniques by which other planetary systems can, in principle, be detected; these techniques will be referred to as either _indirect_ or _direct_ _detection_. Several independent observational methods fall under these general techniques, and we here compare and contrast alternative methods that are being considered for use in detecting other planetary systems.

1. INTRODUCTION

Most hypotheses as to the origin of the solar system, as well as hypotheses regarding the formation of stars, indicate that planetary systems ought to be reasonably prevelant in the galaxy in general and in the solar neighborhood in particular. However, there is presently no unambiguous observational evidence for the existence of any planetary system other than our own. The ambiguous nature of observational studies is primarily due to the fact that the observable consequences of a planetary system are very small, so small as to be beyond the capability of almost any existing instrumentation. The key word here is "existing". Recent NASA-sponsored scientific workshops concerning the detection of other planetary systems (e.g., Black and Brunk, 1980) have shown that there is no fundamental barrier to developing instrumentation that would permit a comprehensive search for planetary systems around stars within ten parsecs of the Sun. The challenge posed by the instrumental aspect of searching for other planetary system, coupled with the long-standing recognition of the scientific import of results from such a search, has led to a recent intensification of effort in this direction.

It will be helpful to give the reader some feeling for the difficulty of the observational problem. One method that could be used to search for other planetary systems is that of astrometry, whereby one makes very accurate measurements of the motion of a star. If a star

M. D. Papagiannis (ed.), Strategies for the Search for Life in the Universe, 167–175.
Copyright © 1980 by D. Reidel Publishing Company.

has dark companions (planets), the apparent motion of the star will be
altered from its motion in the absence of companions. The angular
extent of the alteration in the motion of the Sun due to the Earth is
less than 10^{-6} seconds of arc (as perceived by an astronomer located
ten parsecs from the Sun). That angle is smaller than the angle sub-
tended by an atom in your finger when viewed at arms length! To be
sure, this example is somewhat of an extreme one, albeit a relevant
one to this symposium. By contrast, Jupiter's effect on the Sun as
viewed by our astronomer friend would be 500 μarc-seconds. A typical
astrometric observation (plate) yields relative positional data that is
accurate to a few hundredths of a second of arc. So-called yearly
normal points, involving averages of several tens of nights of data,
are accurate to several thousandths of a second of arc ($\sim$ 5000 μarc-
seconds), an order of magnitude worse than required to detect Jupiter
in the example cited above.

A comprehensive review of the problem of detecting other planetary
systems has been given elsewhere (Black 1980a). The remainder of this
discussion will focus on comparing some of the alternative methods that
are being considered for detecting such systems. Emphasis will be
placed on those methods which were discussed by others at this symposium.
Indirect detection methods are considered in section 2, and direct
detection methods are critiqued in section 3. A brief summary and concluding remarks are given in section 4.

2.   INDIRECT DETECTION METHODS

Any method of detecting planets in other planetary systems by
virtue of some observable effect those planets have on an other object
(e.g., their central star) or physical system (e.g., remnants of the
material from which the planetary system formed) is taken to be an indirect detection method. The three indirect methods that have been
most actively considered are; astrometry, spectroscopy, and photometry.
The same physical effect underlies the first two methods, viz., the
orbital motion of a star about the center of mass (barycenter) of a
hypothetical star-planets system. Astrometric techniques are used to
detect the components of this putative orbital motion that are in the
plane of the sky. Spectroscopic techniques are used to detect the component of this motion that is along the line-of-sight (LOS) between an
observer and the star under study. Astrometry deals with measurements
of angular displacement, and spectroscopy deals with measurements of
Doppler shifted stellar spectral features. The physical effect underlying photometric methods is that if a planet transits the LOS of an
observer to a star, there will be a reduction in the apparent brightness
of that star.

2.1 Astrometry

This method is the one which has been used most often in searches
for other planetary systems. While it has as yet returned no positive
detection, years of observations at many observatories, notably Sproul

Observatory, can be used to establish limits as to what companions nearby stars do not have. The importance of these negative results is not generally appreciated. They suggest that there may be a gap or cutoff in the mass function for members of binary-like systems, and therefore that planetary systems and binary systems do not form by the same process.

As noted above, astrometric studies involve the measurement of the angular displacement of a star from the trajectory it would follow if it had no companions. The amplitude, θ, of this displacement is related to the distance D between observer and the star, and the distance R_* of the star from the barycenter of the system according to

$$\tan \theta = \frac{R_*}{D} \qquad (1)$$

The quantity R_* can be related to characteristics of the unseen planetary companion, viz.,

$$R_* = \frac{m}{M_*} R_p \qquad , \qquad (2)$$

where m and M_* are respectively the mass of the planet and star, and R_p is the separation between the star and planet. (It has been assumed here that $m \ll M_*$ and that the orbits are circular. See Black (1980) for a more general discussion). As $R_* \ll D$, one has that

$$\theta \sim \frac{m}{M_*} \frac{R_p}{D} \qquad \text{radians} \qquad . \qquad (3)$$

Inspection of equation (3) reveals the following general features of this method of detecting other planetary systems: it is most sensitive to detecting massive planets and/or planets which revolve about their central star in large orbits, it is most sensitive to detecting planets revolving around nearby, low-mass stars. As most of the nearby stars have masses less than that of the Sun, the latter feature is a strength of the astrometric method. However, the former feature is a weakness of the method. Kepler's Third Law tells us that the orbital period P of a planet revolving at a distance R_p from a star of mass M_* is proportional to $R_p^{3/2} M_*^{-1/2}$. Bearing in mind that one is likely to need a total observing time that is $\gtrsim$ P to establish the existence of a dark companion, the price for sensitivity gained through large values of R_p is longer total observing time. To the extent that the solar system is a prototype of planetary systems generally, and indeed there is no reason to suppose that it must be, one expects to encounter values of P in the range of one to a few decades. While affording job security, this length of time indicates the need of dedication for both human and instrument.

Given that a companion is discovered by this method, one can learn a fair amount about it in addition to the fact that it exists. For example, the orbital period of the companion will be known from the data, permitting an evaluation of R_p. Knowing R_p, one can determine the mass of the companion (see equation (3)).

During this meeting, we heard about three alternative concepts for
new astrometric systems (see the contributions by Gatewood, Currie, and
Baum for details). Two of these concepts (those discussed by Gatewood
and by Currie) involve ground-based systems, whereas the third concept
involves the Space Telescope. While it is not my intent to critique
these concepts individually, it is worthwhile noting some salient aspects
of each concept.

The type of concept outlined by Gatewood is essentially the natural
extension of classical astrometric instrumentation into the world of
modern technology. Some form of photoelectric array is used to replace
the photographic plate as the detector, and there is no need for an in-
termediate step in the data reduction process (i.e., there is no separate measuring machine). To be fully effective, such a concept would
require a new telescope, one constructed specifically for these types
of observations. As with classical astrometry, this concept would
determine the position of a star with respect to a reference frame
defined by other stars in the field-of-view. The ultimate accuracy of
a system like that outlined by Gatewood is of course unknown, but
theoretically one ought to approach an accuracy $\sim$ 1000 μarc-seconds per
night, and $\sim$ 100 μarc-seconds per yearly normal point. Note that this
level of accuracy would be sufficient to detect Jupiter in the example
cited previously.

The two interferometric concepts discussed by Currie represent a
departure from classical astrometric instrumentation. One of those
concepts involves interferometry (either speckle or amplitude) employing
a single aperture. The other concept involves interferometry using a
multiple-aperture system with moderately long baselines ($\gtrsim$ 10 meters).
These two types of systems would operate in very different ways. The
multiple-aperture systems would operate in much the same way as would
the Gatewood-type system, that is, stellar positions would be determined with respect to a reference frame defined by other stars in the
field-of-view. However, the single-aperture systems would only work
for stars that are separated by relatively small angles (the so-called
isoplanatic angle which is $\lesssim$ 5-7 arc-seconds), because these systems
require coherent behavior in the atmosphere's effect on light transmitted through the atmosphere. This very small effective field-of-view
means that positions cannot be referenced against a frame defined by
other stars; only stars within the isoplanatic patch are useable reference stars. The consequence of this constraint is that it is most
useful in searching for planetary companions to close binary stars.
While this restriction precludes a homogeneous statistical base, it is
the case that close binaries are difficult to study by classical astrometric methods. Thus, the single-aperture systems would be complementary
to their classical counterparts.

Again, one can only estimate the attainable accuracy with these
interferometric systems. Such estimates yield accuracies for single-
aperture systems of 100-10 μarc-seconds, depending on the characteristics
of the binary system under study. Similar estimates for the long-base-

line, multiple-aperture systems yield figures of 100 μarc-seconds.
These accuracy figures pertain to yearly normal points.

The single-aperture systems discussed by Currie are straightforward
extensions of existing methodology. They require no special telescope
(however, a dedicated system is again desirable), and could be operational in the near future. The multiple-aperture systems are more of
an unknown quantity. Studies by Currie, and independently by Shao at
M.I.T., have shown that such systems can work with small ($\lesssim$ 5 meter)
baselines, but their performance and stability at longer baselines
remains to be demonstrated.

The final system considered here is that discussed by Baum, viz.,
the Space Telescope (ST). There are two principal means of doing
astrometry with the ST; using the star trackers employed in the
guidance system of the ST, and using one of the camera systems that
will be on the ST. Although the ST will be free of atmospheric effects,
the simple truth is that the ST was not designed to do accurate astrometry. Thus, while one might hope that the ST would provide a major
breakthrough in accuracy, the accuracy of any of its astrometric
systems is estimated to be only $\sim$ 1000 μarc-seconds. This is worse
than one ought to be able to do from the ground by the time that the
ST is operational. However, a space-based system designed to do astrometry should be capable of accuracies approaching 1 μarc-second.

2.2 Spectroscopy

Spectroscopic observations of stars can be used to detect other
planetary systems, although to date this technique has not been widely
used for this problem. The orbital velocity, V_*, of a star about the
barycenter of our simple planetary system is given by

$$V_* = \frac{Km}{(M_* R_p)^{1/2}} ,$$

(4)

where K is a constant, and the other quantities are as defined above.
The quantity V_* is not necessarily the correct one in terms of a given
observation because the Doppler shift which one hopes to detect is
associated with the LOS component; the speed given by equation (4) is
the maximum possible for the example under consideration. If an observer's LOS forms an angle i with the normal to the orbital plane of the
star, then the Doppler shift is that associated with $V_*(i)$ where

$$V_*(i) = V_* \sin i .$$

(5)

Equations (4) and (5) indicate the following general features of this
detection method: like astrometry it is most sensitive to detecting
massive planets revolving about low-mass stars (although the dependence
on M_* is less than for astrometry), unlike astrometry, it is most sensitive to planets which are close to their central star, and there is no
dependence on the distance D (this is not strictly true because there

is an implicit dependence on D due to the requirement of detecting
enough photons to obtain accurate results). The major point is that
spectroscopic and astrometric techniques are complementary in terms of
the R_p- parameter space.

Until recently, most stellar spectroscopy was accurate to $\sim$ km/s.
Observations done with special care, as well as some studies using
special correlation techniques, have yielded accuracies of $\sim$ 100 m/s.
To put these numbers in perspective, note that the orbital speed of the
Sun induced by Jupiter is $\sim$ 13 m/s. The corresponding number for motion
induced by the Earth is $\sim$ 0.1 m/s. Thus, one needs instrumentation
capable of measurement accuracy in the 1–5 m/s range if spectroscopic
methods are to be competitive alternatives to other detection methods.
One such system (others are also under study and/or development) is that
discussed by Serkowski at this symposium.

A difficulty with spectroscopic detection is the interpretation of
the results. As can be seen from equation (5), the observed effect is
due to a low-mass companion (planet), it may be due to a small value of
sin i. In the absence of independent information on either m or i, one
cannot tell whether a specific star has a planetary companion. This is
not a severe weakness of the method because if there is more than one
detectable period in the data it would argue strongly for a planetary
system rather than a binary viewed nearly normal to its orbital plane.
Also, given that many stars were observed to have small V∗ (i), and
assuming that orbital planes are oriented randomly in space, one could
obtain statistical evidence for the existence of other planetary systems.
The major interpretive uncertainty lies in the fact that we have no
understanding of the intrinsic surface motion of stars at the level of
one to many tens of meters/second, particularily over intervals of time
comparable to likely planetary orbital periods. An effort to monitor
the radial velocity of the Sun at this level of accuracy for a year or
longer would provide very useful clues as to whether this intrinsic
motion will be a problem.

2.3 Photometry

The final indirect method mentioned here is that of photometric
studies. The most comprehensive discussion of this method had been
given by Rosenblatt (1971). The expected apparent dimming, Δm, of a
star due to transit by a planet is given by

$$\Delta m = +2.5 \log\left[1 - (d_p/d_*)^2\right] \text{ magnitudes} \qquad , \qquad (6)$$

where d_p and d_* are respectively the diameters of the planet and star.
In the case of Jupiter transiting the Sun, one has $\Delta m \sim - 0.01$ magnitudes. The analogous value for the Earth transiting the Sun is
$\Delta m \sim - 0.0001$ magnitudes. Currently routine photometric studies can be
done with accuracies $\sim$ 0.01 magnitudes and with care one can obtain
accuracies $\sim$ a few thousandths of a magnitude.

The major difficulty with the method is not accuracy, although there is room for improvement there, it is that viewing a transit event is very unlikely, and one has no idea whether a dimming is due to a planet or some other natural phenomenon. The rareness of the transit event is such that one would need to continuously monitor ∿2000 stars for one year in order to have a high probability of detecting a transit. This large number as based on the assumption that all stars have planetary systems, if only 10 percent of the stars have planetary systems, one must monitor 2×10^4 stars. Because of these difficulties, the photometric method is generally not considered to be a primary search method. Photometry should be done in parallel with either astrometric or spectroscopic searches as it would provide evidence for intrinsic stellar variations that might confuse the other two methods.

3. DIRECT DETECTION METHODS

Any method by which radiation from a planet revolving about another star is detected is defined as a direct detection method. The radiation can be intrinsic thermal radiation, reflected central star radiation, or some form of non-thermal radiation intrinsic to the planet (e.g., the radio bursts emitted by Jupiter). Intrinsic thermal radiation from planets is likely to be primarily infrared radiation, whereas the reflected component is likely to be primarily visual radiation. As for non-thermal planetary radiation, one can only guess as to the relevant portion of the spectrum. If a planet is revolving about another star, the thermal and reflected components of its radiation will be present; the same cannot be said for the non-thermal component. Thus, attempts to detect other planetary systems by virtue of their non-thermal radiation are generally regarded as secondary search methods (this does not mean that they ought not be tried), while attempts to detect either intrinsic or reflected radiation are the principal direct detection search methods.

Very little was said about direct detection methods at this symposium (see contributions to this Proceedings by Owen and by Baum) because direct detection must be done from space. The major technical difficulty associated with direct detection is that one is trying to detect a very dim object (planet) located very close to a bright object (star). Returning to the Jupiter-Sun system as an example, note that Jupiter is some nine orders of magnitude less bright than the Sun in visible light (every two orders of magnitude in brightness corresponds to a magnitude difference of 5 magnitudes) and it is some four orders of magnitude less bright than the Sun in the infrared ($\lambda \sim 30~\mu$). Viewed by our astronomer friend ten parsecs away, the angular separation between Jupiter and the Sun is 0."5.

A detailed general review of direct detection methods has been given elsewhere (Black, 1980a), and there have been specific discussions of detecting intrinsic thermal planetary emission (e.g., Bracewell and MacPhie, 1979; Black, 1980b). Direct detection presents demanding instrumental problems, but they are problems which can, in principle, be overcome with existing or soon to exist technology. The ST will not be able to adequately resolve likely planetary systems, the aperture is too small (by more than a factor of ten) to suppress light outside its Airy disk to a level (10^{-9}) where a planet could be detected. This is not to say that the ST will not discover a planet which is fortuitously bright and around a nearby star (I personally doubt that it will), rather it is to say that the ST cannot be relied upon to conduct a statistically significant survey.

In spite of the difficulties inherent in direct detection methods, there are good reasons to pursue them. They provide information about any planets that are discovered which cannot be obtained by indirect methods, notably the temperature of the planet and possibly some compositional information (see the paper by Owen). Also, direct detection techniques cover a region of a search parameter space that is difficult to cover with indirect detection techniques. Finally, discovery by means of direct detection would in general require less time than discovery by indirect detection methods. The latter require $\gtrsim P$ years, whereas the former would require $\lesssim P/2$ years.

4. SUMMARY AND CONCLUSIONS

Interest in detecting and studying other planetary systems is clearly leading to major advances in instrumentation for astrometry and spectroscopy. These advances include more or less traditional approaches to these two observational methods, plus some new approaches such as those discussed by Currie. Most of the short-term effort is in the area of ground-based, indirect detection methods; there is currently very little effort in the area of space-based, direct detection methods.

An important aspect of any comprehensive effort to detect other planetary systems is the need for complementarity of methods. The detection of other planetary systems is so difficult, and yet so rich in terms of scientific and philosophical return, that we must have a redundancy of not only detection methods (e.g., spectroscopic and astrometric), but also instrumentation within a given method (e.g., more than one astrometric instrument with the requisite accuracy).

Philip Morrison made the comment during those proceedings that, and I paraphrase, not a great deal has changed over the past decade with regard to SETI. If we are able to receive adequate support for planetary detection, Dr. Morrison's remark will not be repeated at a 1990 Symposium on SETI. A search for other planetary systems differs in one important way from a SETI. The former is an interrogation of Nature, and we will be able to state categorically whether a given star has planetary companions, certainly if they are as massive as Jupiter and probably even if they are no more massive than Uranus or Neptune. The challenge was best stated by Dr. Greenstein during a Workshop on Planetary Detection, "We are limited only by our willingness to invest time, thought, and money." I am very excited by the prospect of obtaining firm observational evidence regarding one of the most far-reaching questions posed by mankind.

References

Black, D.C. (1980a)   Space Science Review, 25, pp. 35.
Black, D.C. (1980b)   Submitted to The Ap.J.
Black, D.C. and Brunk, W.E. (1980)   "An Assessment of Ground-Based
  Techniques for Detecting Other Planetary Systems", Vols. I and II,
  NASA CP-2124.
Bracewell, R.N., and MacPhie, R.H. (1979)   Icarus, 38, pp. 136.
Rosenblatt, F. (1971)   Icarus, 14, pp. 71.

THE SEARCH FOR EARLY FORMS OF LIFE IN OTHER PLANETARY SYSTEMS:
FUTURE POSSIBILITIES AFFORDED BY SPECTROSCOPIC TECHNIQUES

Tobias Owen
Department of Earth and Space Sciences, State University of
New York, Stony Brook, New York   11794   USA

Abstract - A consideration of the basic chemistry of life as we know it
suggests good reasons for expecting carbon compounds and water as
fundamental elements in extraterrestrial life. It is then possible to
establish criteria for habitable planets in terms of their sizes and
distances from their stars. If such planets can be found in other
solar systems, and observed separately from their stars, simple
spectrophotometry can reveal whether or not their atmospheres contain
gases such as oxygen, methane and water vapor in concentrations and/or
combinations that would indicate the presence of life.

1.   INTRODUCTION

Detection of life on another planet by spectroscopic observations
requires that the planet can be viewed directly. This has been an
insuperable obstacle in the past, but with the advent of orbiting
space telescopes, it may become possible to distinguish planets from
the stars they orbit. This can be accomplished either with a specially
designed "coronagraphic" instrument or by using a distant occulting
disk, such as the moon. The latter approach seems particularly
promising if the orbiting telescope can be maneuvered sufficiently to
permit relatively long-lasting occultations of the target star. Once
it is possible to obtain an image of an extra-solar planet, it is a
very small step to begin low resolution spectroscopic observations.

We then find ourselves confronting a problem very similar to the
one that existed fifty to sixty years ago, when astronomers first began
examining the other planets in our own solar system with spectrographs,
searching for evidence of life. Our goals are essentially the same as
theirs: we seek for evidence of gases in the atmospheres of these
planets that would suggest the presence of living organisms.

It might be asked why we should spend time trying to detect
primitive forms of life on other planets when (at least in principle!)
our radio telescopes permit us to detect advanced civilizations

M. D. Papagiannis (ed.), Strategies for the Search for Life in the Universe, 177–185.
Copyright © 1980 by D. Reidel Publishing Company.

directly. One reason is that the spectrographic approach is completely general: it also allows us to detect advanced civilizations that are either not signalling or doing so in ways that remain unknown to us. Furthermore, if it should turn out that our civilization on Earth is not an average case but represents instead one of the quickest transitions from atoms to intelligence in the universe, then there may not be enough disk population stars sufficiently older than the sun to permit the existence of many advanced civilizations. In that case, our best hope of finding evidence for extraterrestrial life lies in studies of atmospheres of other planets. The Earth would have provided this clue for at least 2 billion years, more than 10^7 times longer than the radio signals that now reveal the presence of an advanced civilization on our planet.

2.   ALIEN LIFE

What gases should we look for in the atmospheres of other planets? To answer this question, we must confront our bias in favor of life as we know it. The only life we have discovered in the universe so far is life on Earth. We must be careful not to constrain our search for extraterrestrial life unduly by our experiences with this single example. Nevertheless, we can see some reasons why the only life we know is based on carbon chemistry with water as a solvent rather than relying on some other chemical system. Carbon is a highly abundant element and exhibits a remarkable ability to form a host of highly complex molecules. The alternative frequently encountered in speculative models for other forms of life is silicon. The advantages of carbon over silicon are easily illustrated by contrasting CO_2 with SiO_2 and CH_4 with SiH_4. Carbon can move easily between a fully oxidized and a fully reduced state, whereas once silicon combines with oxygen, its tendency to form the large stable polymer known as quartz or some other silicate rock make it very difficult to recover the silicon. Since there is roughly 20 times more oxygen than silicon in the universe, rocks constitute a very significant sink for this element.

At the other extreme, we find that silicon forms very unstable compounds with hydrogen, such that chains with more than three silicon atoms are virtually impossible to construct. Silane itself reacts explosively with oxygen. Hence macroscopic silanogenic life forms visiting Earth would be excellent facsimiles of the fire-breathing dragons reported in Chinese and European legends!

Arguments can also be given in favor of water as life's working fluid. Water is highly abundant, it remains liquid over a broad temperature range, that temperature range is high enough so chemical reactions can occur rapidly, but not so high that large molecules cannot form. Water has a high heat of vaporization and is an excellent solvent. Ammonia is often suggested as a substitute for water and is nearly as good in many categories. But ammonia lacks a subtle property that water possesses, viz., the capability of protecting itself from destruction by ultraviolet light. Water dissociates to form oxygen and ozone which can protect liquid water from further dissociation, provided there is a cold trap in the atmosphere of our hypothetical planet to confine water to lower altitudes. The products of photodissociation of ammonia do not provide this service; nitrogen is totally transparent to photolyzing UV, and other possible products such as methylamine and hydrazine are themselves readily broken down by low energy UV photons. Thus it is problematical whether any planet could have liquid ammonia oceans, unless it has a perpetual cloud cover or orbits a very cool star.

These theoretical considerations find support in what little observational evidence we have at our disposal. We are finding carbon compounds of ever greater complexity in the interstellar medium and in meteorites. We do not find silicones or other mobile silicon polymers; instead we find silicates to be the predominant silicon-containing compounds wherever we look. Similarly, the one object in our solar system that might be expected to have ammonia oceans - Saturn's satellite Titan - is too cold for liquid ammonia to exist on its surface and shows no evidence of ammonia vapor in its atmosphere. Although one might imagine an ammonia greenhouse to operate on Titan, protected by the UV-absorbing haze we find there, this condition has not in fact occurred.

We conclude this discussion by suggesting that it is not a simple accident that life on Earth is based on carbon with water as its liquid. It seems possible that most forms of life elsewhere in the universe will rely on this same chemistry. We certainly do not know enough to exclude the existence of different chemical systems that might operate under very special circumstances and such exotic organisms will simply add to the total value of N. But with our present understanding of the requirements of life, the cosmic abundances of the elements, and the chemistry that occurs in the interstellar medium, in primitive solar systems, and on planetary surfaces, we find some real support for carbon-water chauvinism.

3.   LIKELY ENVIRONMENTS FOR LIFE

Assuming that we are looking for evidence of life as we know it, how do we proceed? As mentioned above, the basic approach is to study planetary atmospheres. We can begin by reviewing the situation in our own solar system. Why is Earth the only planet with an abundance of living organisms?

We can quickly classify the planets in terms of their atmospheres: In the outer solar system we find planets with hydrogen-rich atmospheres; some of these atmospheres may be unfractionated residues from the original solar nebula. These planets are sufficiently large that hydrogen cannot escape from their gravitational fields. Titan offers an interesting exception, being small enough to permit hydrogen to escape. But this satellite of Saturn has such a low surface temperature

(~85°K) that water vapor is absent from the atmosphere and hence there
is no source of oxygen to permit conversion of the primitive reducing
environment. Thus we find methane to be the dominant detected constituent, with traces of other hydrocarbons. In contrast, the small
inner planets have long since lost whatever original hydrogen they may
once have captured from the nebula, and they are warm enough to have
atmospheric water vapor. Indeed, there is no evidence at present to
suggest that these planets ever had significant captured atmospheres;
the amount of atmospheric neon they presently possess is too small.
Thus the atmospheres we now observe in the inner solar system represent
the results of outgassing, escape, and chemical evolution, including
oxidation of major constituents. On both Mars and Venus, we find
atmospheres that are over 95% carbon dioxide. Our present understanding
of atmospheric evolution suggests that our own planet would have a
similar atmosphere in the absence of life and liquid water. A third
type of atmosphere is exhibited by Jupiter's satellite Io, which is so
racked by volcanism that it appears to sustain a tenuous envelope of
SO_2, that varies in density with time and location on the surface.

Finally, we have the Earth, with an atmosphere that clearly announces the presence of life. The huge abundance of free oxygen is
very difficult to explain without invoking biology, since this highly
reactive gas would rapidly combine with the crust and become a trace
constituent if there were not a constant source to maintain its present
level. Moreoever, if all this oxygen were not enough to convince a
distant observer that our planet is inhabited, she could find additional evidence by noting that our atmosphere also contains a trace of
methane. The coexistence of methane and oxygen signifies a large departure from thermodynamic equilibrium; to maintain this state, sources of
both gases are needed. Both are produced biologically: oxygen primarily
by green plant photosynthesis and methane primarily by the metabolism
of anaerobic bacteria living in the guts of grass-eating animals and
the mud of swamps.

If our spectroscopist looked further in the infrared, she could
even find evidence that advanced life forms existed on Earth, since our
atmosphere currently contains detectable traces of fluorocarbons - the
result of leakage from refrigerators and aerosol spray cans. "Advanced"
doesn't necessarily mean "intelligent", as we well know, but the presence
of this artificially produced gas would certainly pique the imagination
of our distant investigator.

In trying to understand why Earth has been so successful as a
nurturent planet, we quickly focus on two unique characteristics: size
and distance from the sun. If Earth were much smaller, it would not
be able to produce and sustain an atmosphere sufficiently massive to
permit the continued existence of liquid water. If it were much larger,
it wouldn't have lost hydrogen so readily, thus preventing the atmosphere
from moving from reducing to oxidizing conditions. The ranges for
these size variations are difficult to specify, although the moon and
Neptune surely represent extreme bounds.

In the case of distance from the sun, we can set similar kinds of boundaries. If Earth changed places with Venus, our planet would experience the "runaway greenhouse" effect that probably led to the virtual disappearance of water from our nearest neighbor: the oceans would heat up because of our greater proximity to the sun, providing more water vapor to the atmosphere, leading to an increased greenhouse effect that would raise the temperature of the oceans further, etc. Finally the oceans would boil and all the water would be in the atmosphere, which would now be so warm that there would be no cold trap to confine the vapor to lower levels where it could be protected from photolysis. Incident ultraviolet sunlight would dissociate the water molecules, allowing the hydrogen to escape and the oxygen to make carbon dioxide and combine with other elements in the planet's crust. In the absence of liquid water and life, the huge amount of carbon dioxide would remain in the atmosphere, providing an efficient green- house effect all by itself. The mean temperature of our planet would approach the extreme presently exhibited by Venus - about 750°K. Life could not exist.

If we traded places with Mars, the outcome is less clear. Since Earth is larger than Mars, our planet has outgassed a much greater abundance of volatiles. It may even have acquired a larger abundance initially. It seems possible that a stable, reasonably temperate environment could be maintained at the distance of Mars from the sun by a planet with an atmosphere consisting mostly of carbon dioxide with a surface pressure equal to or slightly less than the present value on Earth. But much farther from the sun, even this would not be possible and we would be confronting a "runaway refrigerator" that would ulti- mately produce a frozen planet, somewhat analogous to the large icy satellites of Jupiter - Ganymede and Callisto. Thus in our solar system we have a situation that might be described as Goldilocks and the three planets - Venus is too hot, Mars is too cold, and Earth is just right!

While we don't yet have reliable evidence for any solar systems other than our own, we expect that they do exist and that they will resemble ours in their basic properties. Outer planets will be large and hydrogen-rich, inner planets will be small, rocky, with highly evolved secondary atmospheres. Thus we can start our program of searching for other inhabited planets in such systems by looking for inner planets at distances from their central stars (allowing for differences in luminosity between these stars and the sun) comparable to the distance of the Earth from the sun. Somewhat larger distances will be more acceptable than somewhat smaller ones.

The distance of a planet from its star will be relatively easy to determine and will provide the first clue that an Earth-like planet may exist. The approximate size of such a planet can be determined from its apparent brightness and an assumed albedo, providing further discrimination between likely and unlikely environments. But size and distance are only possible clues, they put us firmly in the position of ignorance occupied by astronomers fifty years ago, when many scientists thought (quite reasonably) that both Mars and Venus might be inhabited, since these two planets satisfied contemporary criteria for establishing similarity with the Earth.

4.   SEARCHING FOR LIFE WITH A SPECTROGRAPH

To move beyond this point we must employ our trusty spectrograph. In examining the light reflected by the planet, the dominant absorptions that we detect will be those characteristic of the star itself. But by observing the star directly, it will be a trivial matter to identify these features and to account for their presence in the planetary spectrum. Any additional absorptions that we find will then be caused by gases in the atmosphere of the planet. To find these gases, we must make the customary choices of wavelength range and spectral resolution. Since the planets we shall be examining represent very faint sources ($m_V > 22$), we shall want to use the lowest spectral resolution that is compatible with our goals. We could start with broad-band ($\sim$800 Å) filters, comparable to those employed in the UBVRIJK system. For example, a simple comparison of V and I magnitudes would tell us whether or not the planet's atmosphere contained large amounts of methane. A positive answer would indicate that either our ideas about the similarity of our solar system with others were wrong, since we would have chosen our target planet to be the equivalent of 1 AU from its central star, or that we were looking at a surprisingly prolific, early stage of development of life like ours, or that we had discovered the mature stage of a very different type of life.

Finding no evidence of methane, the next step might be a comparison of the U magnitude with the brightness in the 2000 - 3000 Å range, using a special filter we might call U^*. Finding $U^* \gg U$, i.e., that the planet was very dark below 3000 Å, we would have good presumptive evidence for the presence of ozone. But this would not be definitive evidence, since SO_2 could produce the same effect in relatively small concentrations.

The coup de grace would be a detection of molecular oxygen, which turns out to be a relatively easy experiment. The A-band of O_2 at 7600 Å is remarkably strong, so that very low spectral resolution ($\frac{\lambda}{\Delta\lambda}$ = 20-100) would be able to detect it, assuming (as we are) a total oxygen content comparable to that on Earth (Figure 1). It is difficult to confuse this absorption with other gases, even at low resolution, since other common molecules would have much stronger rotation-vibration bands at longer wavelengths. The special property of O_2 that gives it this distinctive spectrum is its permanent magnetic moment in the ground-state. Without this, O_2 would only exhibit an extremely weak quadrupole spectrum and would be very difficult to detect. We could also search at higher resolution for the Schumann-Runge bands near 2000 Å and the B-band of O_2 at 6800 Å. Detection of either of these would confirm the identification, as would the use of higher spectral resolution that would reveal the band shape at 7600 Å. But these are

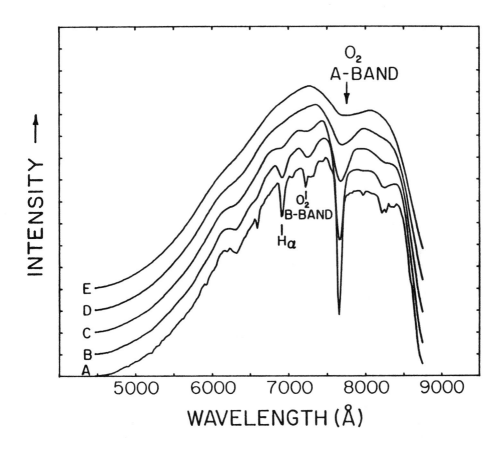

Figure 1: A series of synthetic spectra based on an observation (A) of the sun from the Earth's surface at an air mass of 2.5 and a resolution of 240. The effective resolution of the successive spectra is as follows: (B) – 90, (C) – 40, (D) – 26, (E) – 20. Note that the A-Band absorption is still evident in spectrum (D). The total oxygen column in the line of sight is slightly less than the amount equivalent to an exterior observation of the entire Earth or an Earth-like planet (effective air mass 3.14).

almost unnecessary refinements - absorption at 7600 Å and 2000 - 3000 Å
with a clean spectrum elsewhere would make a very strong case for the
presence of large amounts of oxygen and hence for the existence of life.

There is still more that can be done. This same low resolution
filter spectrophotometry would show whether or not the planet's
atmosphere contained a large amount of water vapor. Here the bands
at 8200 and 9300 Å would be diagnostic. (Much stronger absorptions
are available at longer wavelengths if an infrared detector were
available, but I am specifically tailoring this discussion to the
capabilities of the first Space Telescope simply to show how much can
be accomplished even with a system that is not optimized for the
purpose.) The presence of a large amount of water vapor would imply
the existence of oceans and moderate temperatures. But we would want
an independent measurement of temperature to be sure we were not seeing
a planet in the midst of a "runaway greenhouse" stage, with water
rapidly dissociating to produce the oxygen we observe. It is to
constrain these alternative possibilities that the determination of
the other parameters of the system - the size and albedo of the planet,
its distance from its star - is so important.

Let us return to the situation in which we have detected methane
and no oxygen on our supposedly Earth-like planet. Although we have
no clear record of those times, we can imagine that during the first
billion years of life on Earth, our planet's atmosphere may have passed
through a stage when it contained larger amounts of biogenic methane,
no biogenic oxygen, a large amount of primarily outgassed carbon
dioxide and some fraction of nitrogen. It is not stretching our
imaginations too much to suggest that some other form of life could
develop into a much more advanced stage in a similar atmosphere. The
diagnostic element here is again the non-equilibrium condition - that
both carbon dioxide and methane coexist in an atmosphere requires
some unusual chemistry, and life is a likely candidate. Unfortunately,
carbon dioxide is much more difficult to detect with our simple
spectrophotometric system: we would need to explore the infrared
spectrum in order to discover this gas at low spectral resolution.

But once we have added this infrared capability, we can easily
detect rather small amounts of many other gases, depending how much
of each is present and to what extent their spectral absorptions
overlap. Thus we can look for many other examples of non-equilibrium
chemistry in planetary atmospheres, thereby broadening our search to
include the possibility of life as we do not know it. For example,
we would certainly be able to detect silane if we found our mythical
dragon-inhabited planet!

5.   CONCLUSIONS

We have frequently referred in this conference to the famous Drake equation for determining the number N of advanced civilizations in our galaxy:

$$N = R_\star f_p n_e f_\ell f_i f_c L$$

In this section of the symposium we are asking what we can do to solve this equation without measuring N directly - i.e., without direct contact. The rate of star formation ($R_\star$) is a term that is constantly being redetermined as astronomers improve their under-standing of the stellar populations in the galaxy. A review of modern methods for determining f_p, the fraction of stars with planets is given elsewhere in this volume. I have shown what we need to know to establish n_e, the number of Earth-like planets in a given system, viz., the distance of the planet from the star and its size (plus the spectral type of the star itself!). If we find $n_e > 0$ in a given system by these criteria, we can hope that they imply moderate temperatures, a thick atmosphere, and the presence of liquid water. To evaluate f_ℓ, the fraction of such planets on which life develops, we perform a simple series of spectrographic observations. We first search for large amounts of molecular oxygen - a sure sign that some-thing strange is happening on that planet. We would refine this discovery by searching for water, ozone and methane. With more sophisticated equipment, we could test for other gases that might reveal suspicious departures from chemical equilibrium. Such tests would allow us to consider the possibility of life forms with totally different chemistries from the one we know on Earth, and might even let us go one step further in evaluating the terms in the Drake equation.

If we found a totally bizarre situation - such as the presence of traces of fluorocarbons in our own atmosphere - we might be inclined to invoke the action of intelligent life, thereby assigning a value to f_i. With our own civilization as an example, we would be very surprised if beings that had developed fluorocarbons were not also capable of and interested in interstellar communication, so we would strongly lobby for an intense effort by radio astronomers to search for signals and give us a number for f_c. A new minimum for L would follow once we had made contact.

This is not a bad harvest for a little simple spectroscopy. Furthermore, this approach has the great advantage of achieving all but the last two steps in the absence of any radio signals from the life forms on the target planet, thereby broadening the search for life to include both very primitive and highly advanced manifestations. We would still only be getting data on a few additional stellar systems, but the percentage increase in knowledge over our present situation in which we are trying to extrapolate from a single example would be very large indeed!

P A R T IV

MANIFESTATIONS OF ADVANCED COSMIC CIVILIZATIONS

MANIFESTATIONS OF ADVANCED COSMIC CIVILIZATIONS
AN INTRODUCTION

Sebastian von Hoerner
National Radio Astronomy Observatory
Green Bank, West Virginia 24944

According to detailed but possibly optimistic astronomical estimates (Dole 1970), about half a percent of all stars may be similar
enough to our Sun to have a planet similar enough to our Earth to be
called habitable. The Galaxy then would have a very large number of
about one billion of habitable planets, about 20 light-years away from
each other. Most of them would be very old, about 4 billion years older
than our Earth, because star formation was much more active in the beginning, and because our Sun is not an old star, only half the age of
the older ones.

Regarding the origin and higher development of life on these habitable planets, we know of only one case, our own. Can we do statistics
with n=1? The answer is yes, of course, if we know the rules: n=1
yields an estimator for the average, but none for the mean error. This
means the assumption that we are average has the highest probability of
being right, but we have not the slightest idea of how wrong that is.
This applies to all we know about our origin, past history, and present
state. If we were average, then life would have originated on about a
billion planets of our Galaxy in early times, and would have reached
our present state of intelligence (or lack thereof) already some billion
years ago. And meanwhile, it should have developed far beyond our
present state. Since star formation has only slowed down but never
ceased, newcomers like us may appear on the scene all the time, about
one in 200 years in our Galaxy.

Regarding the further development of the older civilizations, and
especially when it comes to estimating the longevity of a state of mind
comparable to ours, we have no case to go by, not a single one, because
we are mere beginners and no other ones are known. Estimates done without the benefit of data may say something about the author but not much
about the thing to be estimated. Nevertheless, for guiding our SETI
strategies we do need estimates and be it wild guesses. Even a total
uncertainty does not speak against larger SETI efforts. As someone
pointed out in a discussion, Columbus started out with considerable expenditures for an impossible goal under wrong assumptions, but he still
discovered America.

M. D. Papagiannis (ed.), Strategies for the Search for Life in the Universe, 189–196.
Copyright © 1980 by D. Reidel Publishing Company.

The main questions of Chapter IV are the following two:

What activities should we expect, visible to us?
Why don't we see any?

As for the answers, all we can do is try to generalize what we know
about our present state, as well as what we guess about our future,
allowing in both cases for a wide (unknown) scatter of "their" various
developments, desires, and ways of life. On the other side, it seems
a good convention to set a constraint: not to consider activities
which would violate any laws of physics as we know them at present.
But within that constraint we should allow for a highly advanced technology in general, and for large-scale great enterprises in at lease a
few cases.

We should keep in mind that it is especially these few cases in
which we are mostly interested, out of the possibly very large number
of different civilizations developing during very long times. Freeman
Dyson has put forward the statement: Whatever <u>can</u> be done (with the
available matter and energy, within the laws of physics), <u>will</u> be done
(given enough time and space). I like to call this the "ergodic theorem
of technology", and I feel strongly its convincing power. Even in the
present absence of evidence.

What about the only case we know? We must admit that our own
activity at present is mostly self-destructive. Our frightening arms
race has completely gotten out of hand and cannot continue for long.
Either we get more reasonable or we blow each other to pieces. The extent of this activity is difficult to visualize, and a few rough numbers
might help. All the big nations spend 1/10 of their gross national
product on their armed forces and on the development and manufacture of
bigger and better weapons. The USA alone spends roughly 150 billion
dollars per year, about the same as the USSR, adding up to a world-wide
total of 400 billion dollars per year for what each one calls "defense",
a lot more than is ever spent on health, education or science. For comparison, the budget of NASA is 4 (that of NSF is only 1) billion dollars
per year. Furthermore, in 1976 the underdeveloped nations received 13
billion dollars of economic help from the industrialized nations, but
bought weapons from them for 18 billion dollars (Sampson 1977) which
means that "economic help" is a euphemism for a large-scale subsidy of
the big nations for their own military industry, on top of their of-
ficially admitted defense budget. Finally, the accumulated destructive
power of all existing nuclear bombs was already in 1972 about 40,000
Megatons of TNT, which means 10 tons of dynamite for <u>each</u> living person
on Earth (von Hoerner 1975). If we were visited by the famous Little
Green Men from outer space, we might find it difficult to explain why
people intelligent enough to build nuclear bombs could really be stupid
enough to do so.

These severe problems are quite relevant to SETI, in many respects.
First of all, we cannot do any long-term projects if we do not stay

alive. Second, it could well be that these pathological looking problems and their threatening consequences are the general rule, that any
species which develops fast to become dominant on its planet is necessarily aggressive, and that the destructive ability of technology then
leads soon to a self-elimination of aggression (von Hoerner 1975).
Which would imply a self-elimination of higher developed life in general
or at least of its technical varieties. Third, the ergodic theorem
means, even so, that at least some of the many hapless "intelligent"
critters become reasonable enough to make it, to harness the tremendous
powers of technology along creative, not destructive, ways into the
future. Fourth, the very numbers quoted above prove clearly that extremely large amounts of money, brain and energy can be directed to
world-wide goals which seem important. And our present use of these
large amounts proves equally clearly that the choice of goals which
seem important is not necessarily dictated by sober reasons like cost-
effectiveness, careful use of energy and dwindling resources, improvements of living standard and safety, or the like.

Finally, if we seriously want to survive (seriously enough to replace words by actions), we must stop the deadly arms race and must
greatly reduce our armed forces, which means that the extremely large
amounts of money, brain and effort just mentioned must be redirected
to new goals, and a beginning of "astro-engineering" might come in
quite handy and naturally (von Hoerner 1978). Maybe this is a general
development of those civilizations who overcome their heroic but suicidal
aggression: starting exploration and exploitation of the resources of
their planetary systems, and trying large SETI projects as well.
Couldn't this be felt as a greater challenge than killing each other?
And if our planet Earth is getting too small for our ever-increasing
power struggles and expansionistic drives, why on Earth should we stay?

Large self-sustaining and profitable colonies throughout our solar
system can be constructed and maintained even with our beginner's technology of today, and even with less money than our arms race. After an
original start from Earth, the colonies could use their own resources:
minerals from the Moon and the asteroids, and plenty of solar energy.
They could grow and multiply on their own. A good deal of our heavy
industry could be moved to space, leaving the Earth a more enjoyable
place. And large-scale mining operations, especially on the thousands
of asteroids, could supply us with the metals and minerals getting rare
here on Earth. We have begun to feel badly the "Limits to Growth", and
if we want to be less limited, we ought to grow out of our terrestrial
cage.

That these are real possibilities for us, not just science fiction,
has been well demonstrated by many physicists and engineers. A good
general summary is given by one of the main pioneers, Gerard O'Neill
(1977), in his book "The High Frontier; Human Colonies in Space." And
there are two volumes "Space Manufacturing Facilities; Space Colonies"
of the American Institute for Aeronautics and Astronautics, with 76
technical and engineering papers about the various aspects (Grey, 1977).

Every year there is an International Astronautical Congress, and I
attended the 28th one, 1977 in Prague. A total of 379 papers were presented, and the following breakdown into several categories is quite
interesting (even the Law has carved itself a large niche out of empty
space):

Number of Papers, at 28th International Astronautical Congress

Present application and handling	Science, industry, future
43 Communication satellites	28 Scientific spacecraft
54 Earth, oceans, sensing	26 Medicine, biology
60 Space law	21 Space-based industry
43 Orbits	16 Large structures in space
22 Propulsion, power	11 Solar system exploration
22 Others	22 Space mining, costs + benefits
244	11 SETI
	135

What, then, are the manifestations of our activities, possibly
visible to others? At present, these are the radio signals from our
TV stations and military radars; Dr. Sullivan will give a report on
this subject, showing that our presence could be detected in the immediate interstellar neighborhood, even by our own means of detection.
Regarding our nearer future (if we have any), with many colonies and
mining industries, the numerous and powerful radio links for communication and navigation would probably be a good deal stronger, reaching
somewhat further. Thus, if we had interstellar neighbors similar to
us and not too far away, a few hundred light-years, say, we could detect them if we built a large antenna system like Project Cyclops
(Oliver 1971) for eavesdropping on their local or interplanetary broadcast.

In addition to local broadcast there may be much more powerful
signals to search for. If the older civilizations have joined together
in a large community, Bracewell's "Galactic Club" (1974), they might
be interested in attracting the attention of newcomers and in providing
them with some higher education. There may be radio beacons, transmitting at special frequencies which can be guessed by the newcomer.
The long waiting time for answers, caused by the finite speed of light,
is no argument against interstellar communication -- even if it would
imply one-way communication only. Our whole wealth of tradition,

culture, and learning has been given from one century to the next in terms of one-way communication. Thus, we may search for signals meant to attract our attention. But maybe we would need really large-scale dedicated efforts before becoming successful.

The space activities mentioned for our nearer future may already be called astro-engineering, though still on a small planetary scale. The next step to interstellar distances might be sending unmanned exploratory probes to prospective nearby stars. One such possibility has been worked out in some detail: "Project Daedalus" (Martin 1978). It will be described by Dr. Martin. Thus, if the others have had similar ideas and enough zest for their realization, we should look for interstellar probes exploring us, as Ron Bracewell has suggested many times. It is very unfortunate that Dr. Kardashev could not give a paper on "Astro-engineering" as planned.

Let us assume we have proceeded in exploring and colonizing our own planetary system, a few hundred years from now. What future steps may we anticipate? Two different, though not exclusive, possibilities have been suggested. If we want to exploit our own system to the full extent, using whatever matter and energy is provided, we may take a large planet apart and build with its matter a broad ring of flat islands, coasting in a circle around the Sun, somewhat outside the Earth's orbit, as suggested by Freeman Dyson (1959). This would multiply our living space by a factor of 10^8. All the energy supply is solar power, and the waste energy must be released as infrared radiation. Thus, we should search for celestial infrared sources. This has been done, but so far all detected sources have found a natural explanation, for example, as processes of star formation.

Regarding the other possibility, suggested by Hart (1975), Jones (1976), Papagiannis (1978) and others (von Hoerner 1978), let us try to visualize the future life in our colonized planetary system. These colonies will grow with their own babies and grandchildren, they will multiply on their own in numbers and activities, coasting in space far away from Earth (and its bureaucracy and power struggle). The cultural and emotional ties to Earth will gradually decrease with each following generation, until finally some colonies may break off completely after some Declaration of Independence, to start out on long interstellar voyages in their "mobile homes", without needing any drastic change in their way of life. The voyages will last many generations, and if we drop the old prejudice that space travel must be completed within an individual's lifetime, interstellar distances can be traveled even with our beginner's technology. We could do it now with 1/100 the speed of light, and probably with c/10 after some more advances in technology.

These travelers may stop on their way at promising places, for mining purpose and for colonizing nice inviting planets. They might stay on a planet and slowly multiply and colonize the whole system again, until after a certain "settling time" some of them start out again in new mobile homes for further voyages. In this way, a wave of stepwise colonizations would be started, covering more distance and volume all the time, finally colonizing whatever looks habitable throughout our whole Galaxy.

As an example, we assume an average trip with a speed of $c/30$, to cover a distance of 30 light-years, which then takes 1000 years or 30 generations. In order to cover 100,000 light-years, the diameter of our Galaxy, 3000 such consecutive trips are needed, with a total traveling time of three million years. If a mobile home houses 10,000 people, and if they multiply on a planet by 2% per year as we do now, then it takes only 700 years of settling time until their number is 10^{10} again and another next step may be started. The 3000 consecutive settlings then take two million years, about the same as the traveling. In total, we could colonize our whole Galaxy, from one end to the other and every nice planet in it, within five million years. This is a very short time for astronomy, and not too long for biology either.

Frank Drake raised the question of whether something so high in energy-consumption and so low in cost-effectiveness as interstellar travel could ever be funded by a reasonable government. For an answer, I would like to mention again the 40,000 Megatons TNT of our nuclear bombs which did exist in 1972 (von Hoerner 1975), and which may have increased to 100,000 Megatons by now (12% annual increase assumed). Since TNT has 4×10^{10} erg per gram, the energy available in our bombs is at present 4×10^{27} erg. Freeman Dyson (1966, 1968) has described "Project Orion", the design of an interstellar space ship whose propulsion is done by exploding nuclear bombs far behind, catching the energy with a large shield and a long shock absorber, all of which is quite possible with our present technology. If we want a travel velocity 1/30 the speed of light, and if we need 10 tons of ship's mass per passenger, then the energy which our governments have already funded for bombs would do for almost 1000 interstellar passengers in a space ship of 10,000 tons. This energy which is stored in our bombs has already been provided by our present primitive earth-bound technology, whereas a future advanced planetary industry could provide a large multiple of that. Furthermore, let us keep in mind that it is not the sober bare necessities of life which make it worthwhile and attractive; it is the luxuries and great adventures, and they are never governed by cost-effectiveness, they tend to go right to the limit of whatever is possible.

What, then, are the manifestations to be expected? Any one of the many early civilizations could have started a wave of galactic colonization, and according to the ergodic theorem at least a few should actually have done so. It sounds like a very natural, straightforward proposition. Just like hundreds of islands spread out over thousands of miles of the Pacific Ocean were colonized by Polynesians long ago, island-hopping in their small fragile boats.

In general, life shows the tendency to expand, to fill every

possible niche, from deserts to polar regions. Life has started in the
water, has conquered first the land and then the air; it begins now to
conquer empty space nearby, and it may naturally proceed to conquer
interstellar distances, to inhabit whatever is (or can be made) habitable. Furthermore, the milestones of evolution are set by introducing
new means of information handling or language. Self-reproductive life
began with the genetic code, higher life with a nervous system, human
culture with speech; and all this should naturally proceed to the
Galactic Club. We then have the great puzzle that the whole Galaxy
should be teeming with life, obvious in many ways, and our Earth should
have been colonized long ago, whereas we have not yet seen any evidence
of extraterrestrials, and we are certainly not the descendants of any
early settlers.

Possibilities of advanced technology have been discussed which are
even more dramatic than just colonizing. Kardashev (1964), for example,
suggested that advanced technologies will be mainly defined by their
supply of energy: Type I uses the energy of a planet (as we do now);
Type II uses the energy production of a star (factor of 10^{13} greater);
and Type III may use the energy output of a whole galaxy (another factor
of 10^{11}). Since the transition time from one type to the next is probably relatively short, we may expect to find civilizations of all three
types but not much in between. We may not be able to guess the activities and manifestations of Type II and III, but we hope we would recognize them as something "artificial" whenever we see them.

Why, then, has Earth not been colonized, why are no extraterrestrials mining in our planetary system, why have we not received any
radio signals, and why do we not see any evidence of astro-engineering?
Does it mean we are alone?

Many reasons against colonizing and astro-engineering have been
discussed (von Hoerner 1975, 1978): self-destruction of technologies,
biological degeneration, stagnation by over-stabilization, change of
cultural interests, the zoo-hypothesis, space technology not becoming
cost-effective, a settling time longer than a million years, colonization only as a random walk. All of these reasons may hold in some
cases, maybe in most cases, but hardly with no exception at all in a
very large number of cases.

There are several possible conclusions. It could be that one in a
billion of habitable planets, Earth, has been overlooked by chance or
neglected on purpose by the colonizers. Regarding evidence of astro-
engineering, maybe we have not yet looked for the right thing or we do
not understand what we see; maybe ground-based astronomy is too limited,
and observations from space have just barely begun and we may see
"artificial" objects if we continue on a larger scale. And our searches
for radio beacons have only been very sporadic and limited, and success
may come only after a large dedicated effort like Project Cyclops or
something similar. On the other side, we may actually be alone, for
some reason not yet understood; and then it is up to us to go and colonize the Galaxy.

Thus, let us finally start some large dedicated SETI searches of various kinds. And then, if we cannot find a Galactic Club, let us go and found one.

REFERENCES

Bracewell, R. N., 1974. "The Galactic Club", Portable Stanford.
Dole, S. H., 1970. "Habitable Planets for Man", Elsevier, New York.
Dyson, F. J., 1959. Science, 131, 1667.
Dyson, F. J., 1966. "Perspectives in Modern Physics. Thoughts on the Search for Extraterrestrial Technology", Interscience, New York.
Dyson, F. J., 1968. Physics Today, 21, 41.
Grey, Z. (ed.), 1977. "Space Manufacturing Facilities, Space Colonies", Vol. I and II, American Institute of Aeronautics and Astronautics, New York.
Hart, M. H., 1975. Quart. J. Roy. Astr. Soc., 16, 128.
Hoerner, S. von, 1975. Journal of the British Interplanetary Society, 28, 691.
Hoerner, S. von, 1978. Naturwissenschaften, 65, 553.
Jones, E. M., 1976. Icarus, 28, 421.
Kardashev, N. S., 1964. Soviet Astronomy, A. J., 8, 217.
Martin, A. R. (ed.), 1978. "Project Daedalus", Supplement of the Journal of the British Interplanetary Society.
Oliver, B. M., 1971. "Project Cyclops", NASA/Ames Research Center.
O'Neill, G. K., 1977. "The High Frontier, Human Colonies in Space", William Morrow.
Papagiannis, M.D.: 1978, Origin of Life, ed. H. Noda, Center Acad. Publ., Tokyo, Japan, p. 583.
Sampson, A., 1977. "The Arms Bazaar", Viking Press, New York.

STARSHIPS AND THEIR DETECTABILITY

Anthony R Martin and Alan Bond*
Culham Laboratory, Abingdon
Oxon, OX14 3DB, England

ABSTRACT

The feasibility of interstellar vehicles journeying among the
stars is considered. The starting point is the results of Project
Daedalus, which demonstrated that a relatively crude interstellar
probe would be possible with only modest extrapolations of present-
day science and technology. The discussion is then extended to
consideration of fast starships, relatively small vehicles which
travel at a reasonable percentage of the speed of light, and much
slower, much larger world ships which spend millenia on their journey.
Some of the characteristics of such vehicles are commented upon, in
a more general examination of the impact that the existence of starships has upon the detectability of advanced technical civilisations
in the Galaxy.

1. INTRODUCTION

Ever since the paper by Cocconi and Morrison in 1959, the main
strategy followed in the Search for Extraterrestrial Intelligence
(SETI) has been to assume that if advanced technological civilisations exist elsewhere in the Galaxy, then they will be attempting to
enter into radio contact with other similar civilisations. Thus,
several searches have been made of the radio sky in attempts to detect non-natural signals, and several design studies and programme
plans have been presented involving the construction and use of dedicated facilities to be used to detect such signals.

An alternative strategy, upon which less emphasis has been
placed, is that of looking for signs of environmental modification
produced by advanced civilisations who will, it is presumed, command
and use much greater energy resources than our planet has at its

---

* The authors would like to note that this work is a private venture
  and is in no way connected with their duties at Culham Laboratory.

M. D. Papagiannis (ed.), Strategies for the Search for Life in the Universe, 197–226.
Copyright © 1980 by D. Reidel Publishing Company.

disposal at the present time. These are the so-called Kardashev civilisations.

The third strategy proposed, and one almost totally neglected by the scientific community at large until relatively recently, is the attempt to detect phenomena that would be characteristic of an interstellar craft making a journey among the stars - a starship in passage.

Many discussions of search strategies in the past have tended to argue that starships are impractical, giving generally vague reasons or biased calculations to back this up. To many the concept of a starship verges on the heretical, at the worst, or lies in the domain of science fiction, at the best, and such arguments have tended to become accepted too easily. It is our contention that starships are not impractical in certain cases, and that realistic scenarios can be conceived in which certain classes of vehicle fall naturally into place. We use this as a basis for a discussion of the impact upon the detectability of advanced civilisations in the Galaxy.

The literature on interstellar travel and communication is extensive. It is not our intention to produce a review of all aspects of the debate - to do so would require more space than we have and detract form the main point of our paper. We will not, therefore, give detailed reference to the work upon which we have based our discussion, but instead refer the interested reader to the comprehensive Bibliography compiled by Mallove, Forward, Paprotny and Lehmann, to be published in the Journal of the British Interplanetary Society (JBIS, $\underline{33}$, 201, 1980) to coincide with the appearance of the present paper.

As a starting point for our discussion we must first consider whether a starship can be built, or not. To answer this question in the affirmative we rely entirely on the results of Project Daedalus - a detailed engineering and scientific study of all aspects of a simple interstellar mission.

2. PROJECT DAEDALUS

2.1 Introduction

During the 1960s and early 1970s gradual progress was being made in the area of pulsed power technology, using lasers or relativistic electron beam generators. This technology is directed at producing energetic beams of photons or electrons in bursts, with energies measured in kilojoules or more, lasting for several hundred nanoseconds or less. That is, pulses of power at levels greater than 10^{10} watts are produced.

It became apparent in the early 1970s that this technology could
be employed to achieve the compression and heating of small quantities
of thermonuclear fuels to conditions of temperature and density such
that nuclear fusion reactions would proceed at such a rate that more
energy would be liberated than consumed by the process of ignition.
Under such circumstances, high energy plasmas would be produced, and
it was realised that these may be employed for space propulsion purposes, with the prospect of producing exhaust velocities of up to
several million metres per second, yet still achieving reasonably
low system masses.

With these developments in mind, the idea of forming a Study
Group, under the wing of the British Interplanetary Society, to in-
vestigate the feasibility of a simple interstellar mission was discussed by several members of the BIS in 1972, and the Project was
inaugurated in January 1973. The final report on Project Daedalus,
published by the BIS in 1978, represents the spare-time effort of a
core of 13 professional scientists and engineers working over a period
of five years. It is conservatively estimated that at least 10,000
man-hours went directly into work for the Project.

The major objective of Project Daedalus was to carry out a feasibility study for a simple interstellar mission, using only present-
day technology and REASONABLE extrapolation for near-future capabilities. It was not intended to produce the "blue-prints for a starship",
but rather to establish whether any form of interstellar space flight
could be discussed, in sensible terms, within established science and
technology. The area of propulsion was known to be the major limiting
factor for the mission and from the outset of the study a mission
consisting of a simple, undecelerated, unmanned stellar flyby was
believed to be the only one worth serious consideration.

It was planned that the study would involve the examination of all
aspects of the mission, in an integrated manner, to determine whether
it was realistic, to establish some of the important interactions
that occur with such a vehicle, and to spotlight problem areas or
areas where knowledge is lacking. The results of the study, and the
proposals and designs presented, are not intended to be the final word
by any means. They represent a possible guideline to the type of system and the performance which could be envisaged, using currently
available ideas and methods. The study may be aptly described as "an
attempt to prove an existence theorem".

In assessing the success of the report in achieving the aims of
the Project, one fact must be borne in mind. Whenever an area of the
study could be satisfied with conventional, or even pedestrian, technology, then that solution was adopted. However, if the technology
was found to be currently lacking, then it was projected, it is hoped
realistically, within reasonable bounds and within the timescale en-
compassing the remaining years of this century. This philosophy of
approach to the Project has resulted in apparent contradictions of

technological level on the vehicle derived. Thus the structure, communications and much of the payload are entirely within present-day capabilities, whereas many details of the propulsion system, the machine intelligence, and the adaptive repair systems on the vehicle are several decades away. It is hoped that in the latter case capabilities for the future have not been overpredicted. In the former case they have certainly been under-predicted.

The results of the study indicated that if we can realise the technological steps in nuclear fusion and electronics, predicted by reasonable extrapolation of today's technology, then interstellar missions of a limited type will be possible. The main difficulty would be the provision of sufficient quantities of helium 3, and this must await developments in Solar System exploitation, thereby permitting its extraction from the outer planets, notably Jupiter. Such a capability will not be available for many decades, and so places a very approximate lower limit to the time at which such missions would become possible, probably the latter part of the 21st century. Of course, the development of the nuclear pulse rocket itself will have a great impact on interplanetary flight and may hasten the process of exploitation.

2.2 Mission Profile

The mission profile which was studied in detail was that proposed at the beginning of the study, namely an unmanned flyby of Barnard's Star at a distance of approximately six light years from the Sun. The vehicle which eventually emerged from the analysis consisted of a two stage nuclear pulse rocket capable of traversing the distance in about 50 years. The two stage concept arose from the failure of early studies of single stage vehicles, the main problem being erosion by the interstellar medium of the large engine of the vehicle. Thus the addition of a smaller engine and redistribution of the propellant was a logical step.

The design, shown in Figures 1 and 2, calls for a vehicle having a mass at engine ignition of about 54,000 tonnes, of which 50,000 tonnes would be propellants. The total mission would involve about 20 years design, manufacture and vehicle checkout, 50 years flight time and six to nine years transmittal of information back to the Solar System. Therefore, it appears that even the simplest interstellar missions will require a commitment of funding for 75 to 80 years.

The vehicle would leave the Solar System probably beginning in near-Jovian space. The boost period (covering three propellant tank drops and a stage separation) would last for 3.8 years, after which a coast velocity of about 12 to 13 percent of the speed of light would be attained.

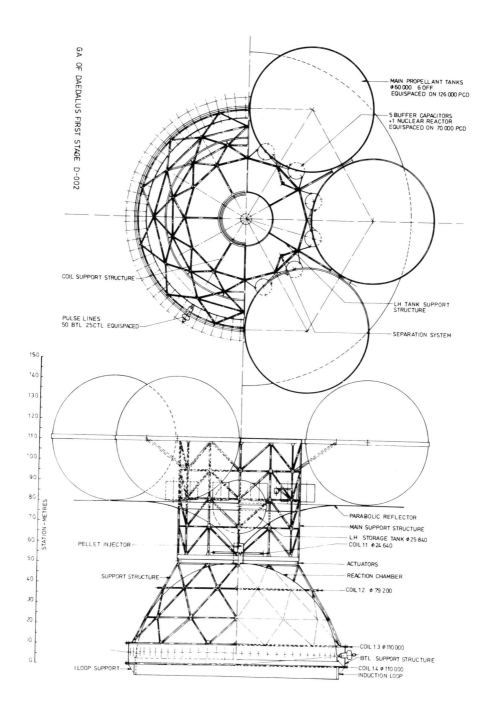

Figure 1. Daedalus vehicle first stage configuration

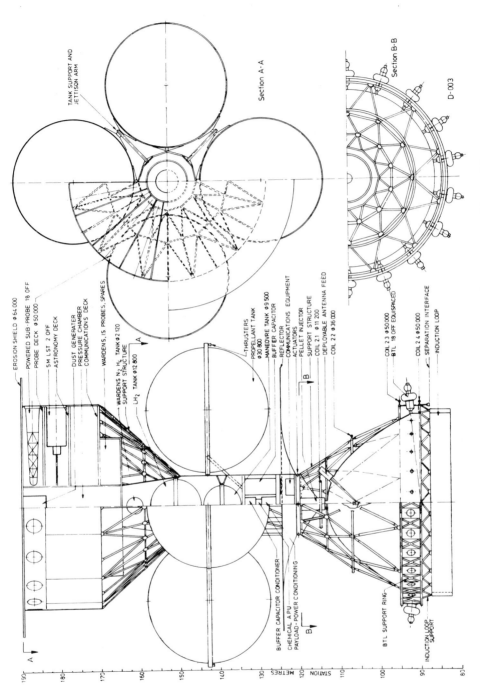

EROSION SHIELD φ 64 000
POWERED SUB-PROBE 18 OFF
PROBE DECK φ 50 000
ASTRONOMY DECK
DUST GENERATER
PRESSURE CHAMBER
COMMUNICATIONS DECK
5M LST 2 OFF
WARDENS, IS PROBES, SPARES
WARDENS N₂ H₂ TANK φ 2 120
SUPPORT STRUCTURE
LH₂ TANK φ 12 800

TANK SUPPORT AND
JETTISON ARM

Section A-A

THRUSTERS
PROPELLANT TANK
φ 30 600
MANEUVRE TANK φ 9 500
BUFFER CAPACITOR
REFLECTOR
COMMUNICATIONS EQUIPMENT
ACTUATORS
PELLET INJECTOR
SUPPORT STRUCTURE
COIL 2.1 φ 11 200
DEPLOYABLE ANTENNA FEED
COIL 2.2 φ 36 000

COIL 2.3 φ 50 000
BTL 18 OFF EQUISPACED

COIL 2.4 φ 50 000
SEPARATION INTERFACE
INDUCTION LOOP

Section B-B

D-003

BUFFER CAPACITOR CONDITIONER
CHEMICAL APU
PAYLOAD - POWER CONDITIONING

BTL SUPPORT RING

INDUCTION LOOP
SUPPORT

STATION METRES

Figure 2.   Daedalus vehicle second stage configuration

During the coast the payload would remain active measuring various parameters associated with the interstellar environment. For the actual encounter a dispersable payload was evolved, which would be deployed rather like the payload of a MIRVed missile, using the main propulsion system for the manoeuvres. Following the encounter the information gathered would be transmitted back to the Solar System by microwave. The payload to do all of this would have a mass of about 500 tonnes, a large fraction of which would be in the dispersable sub-payloads.

2.3 Propulsion System Principles

Before the study began it had been established that the only high performance propulsion system having the potential for interstellar flight within the scope of predictable 20th century technology was the nuclear pulse rocket with external ignition. The design of this type of engine in its earlier stages was influenced by the work of Winterberg and other authors who had discussed the application of such an engine to missions within the Solar System.

Initial estimates suggested that it would be necessary for the engine hardware mass to have a weight allocation of about 500 tonnes, irrespective of the actual size of the engine, and this remained relatively constant throughout the study. A specific power of 100 MW/kg and an effective exhaust velocity of 10^7 m/sec were the original design estimates, and this exhaust velocity was used as one of the design parameters, as it was felt to be one of the most important characteristics of the engine.

The design of the engine proved a taxing problem, but it was found that the scale effects due to the large size of the vehicle helped. Previous nuclear pulse designs had used deuterium and tritium as a propellant

$$D + T \longrightarrow {}^4He + n + 28.2 \times 10^{-13} \text{ joules,}$$

but it was realised that the neutrons produced from the fusion reaction would lead to unacceptable mass requirements for shielding for the structure and payload and to large waste heat rejection systems. The reactions between two deuterium nuclei also produces neutrons

$$D + D \left\langle \begin{array}{l} {}^3He + n + 5.2 \times 10^{-13} \text{ joules} \\ T + p + 6.4 \times 10^{-13} \text{ joules,} \end{array} \right.$$

and so the Daedalus vehicle uses deuterium and helium 3 fusion, as the reaction products here are all charged particles and contribute to the thrust from the engine

$$D + {}^3He \longrightarrow {}^4He + p + 29.4 \times 10^{-13} \text{ joules.}$$

Small spheres, of a few centimetres in diameter, of deuterium and helium 3 are injected into the centre of a cusp-shaped magentic field, and as they reach the target point they are hit simultaneously by high-power electron beams. The outer layers of the spheres are ablated away due to the high heating rates and energy deposition in these outer regions. This ablation results in very high surface pressures being generated and the fuel is compressed and shock heated, with the central regions attaining temperatures at which thermonuclear reactions can occur.

The resulting expanding plasma ball is highly conductive and sweeps aside the magnetic field in the reaction chamber; surrounding the reaction region is a very thin but very large diameter molybdenum shell. Due to the rapidity of the deformation of the magnetic field it would effectively be confined and compressed between the conducting shell and the plasma ball. The kinetic energy of the plasma would be temporarily stored in the magnetic field, which would then reverse the motion of the plasma and eject it at high velocity along the axis of the engine. The momentum of the exhaust would be transferred to the molybdenum shell during this process and subsequently, by means of a thrust structure, to the vehicle.

The energy required to ignite the following sphere would be extracted from the current reacting pellet VIA an induction loop situated at the reaction chamber exit. This loop would charge a large number of Blumlein-type transmission lines to a potential of several hundred megavolts. The stored electrical energy would be subsequently discharged through field emission brushes and a plasma bridge into the next sphere.

The propellant would be carried as preformed spheres in several drop tanks. The spheres would be basically a deuterium honeycomb filled with helium 3, and the storage temperature would be about $3^\circ K$. The size of the individual spheres is basically limited by the weight of the ignition system needed. A larger size of sphere requires a heavier system, and the present design represents a partial optimisation of explosion repetition rate, set a 250 per second in the design, against system weight.

Because mid-course correction burns and considerable terminal propulsion capability for distributing the payload packages near the target will be required, it is necessary to have a relatively large engine available throughout the mission. This is achieved by carrying the second stage engine, which has a smaller overall diameter and a smaller thrust than the first stage engine, throughout the mission.

2.4 Propulsion System Engineering and Auxiliary Power

In order to keep the mass of the propulsion system low it was necessary to optimise the ignition systems on each engine. The final result was a large number of pulse forming transmission lines directly charged from the engine induction loop to very high voltage. On both stages the potential came out to be in the region of 250 million volts, with a maximum field gradient of 10^9 volts/metre. These values are outside present technology, but it is thought that granted the clean manufacturing environment of space, and very pure materials, they should not be impossible values to work at.

The high fuel burn-up fractions required of the Daedalus engines would also require very sophisticated diode design for the electron beam generators, involving varying current density and magnitude during the pulse. Again, although no one has so far achieved this behaviour it does not appear an impractical objective.

The fuel pellets need to be injected into the reaction chamber at high velocity in order to reach the target point at the correct time. In the case of the first stage motor, this velocity is 13.75 km/s. It is proposed to achieve this by providing each pellet with a thin superconducting shell and accelerating them into the chamber with an electromagnetic gun. This would simply involve an array of coils and capacitors producing a travelling magnetic wave on which the pellet would be accelerated. Such a device does not appear to be much beyond current technology.

Nuclear radiation from the engines is expected to be very low (although much beyond human tolerance!) This is due to the choice of Helium 3 as one of the fuels. The DHe^3 reaction produces very few neutrons anyway, but even those that are produced are mostly captured in the He^3 at the reaction conditions considered here.

The reaction chamber design followed from a policy of keeping cooling systems as simple as possible. It is completely passively cooled by radiation cooling. It will, however, be a very high temperature component, being heated by eddy currents during the engine operation. The thermal radiation from this item does give other cooling problems, particularly with the electron beam generators which have to be refrigerated to keep them sufficiently cool to prevent electrical breakdown. The reaction chamber would be of molybdenum and have a double-walled construction with cross ties and internal helium pressurisation. The function of this construction is to rigidise the very thin shell and prevent damage which would otherwise occur if a pellet failed to ignite (a condition which tends to turn the chamber inside out).

A summary of calculated engine parameters is given in Table 1.

The auxiliary on-board power for the vehicle would be provided by four nuclear fission reactors, having two main levels of power output. During the coast phase, the power supply would be 270 kW to power the attitude control systems, cooling systems, payload, computer and low data rate communications. However, during the encounter phase and the post encounter, the reactors would operate at a power

level of about 2.5 MW for a period of up to three years, while all the
accumulated data were transmitted home. Waste heat from the auxiliary
power supply is dumped overboard VIA four waste heat radiators.

In order that short duration, high power level activities can be
carried out during the coast period, e.g. high thrust attitude control
pulses, a high voltage 10 MJ capacitor is provided which is recharged
slowly VIA the reactors. During the boost period, however, this
capacitor is charged from the main engine induction loop, since demands on it are greater in that period.

Table 1. Some Main Engine Parameters.

Parameter	1st Stage	2nd Stage
Flow (kg/s)	0.711	0.072
Power (W)	4.38×10^{13}	3.36×10^{12}
Chamber diameter (m)	100	44
Chamber mass (tonnes)	275	31
Field coil mass (tonnes)	194	73
Igniter mass (tonnes	425	125
Exhaust velocity (m/s)	1.06×10^7	9.21×10^6

2.5   Vehicle Structure

The structure of the Daedalus vehicle may be sub-divided into
several parts. These are, for the first and second stages: reaction
chambers; reaction chamber thrust structure; core and fuel tanks.
The payload bay is located at the forward end of the second stage,
and has a series of four decks having sub-probes, scientific instrumentation, communications equipment, and wardens.

The thrust structure, which carries the loads from the reaction
chambers, employs pin ended struts to reduce thermal stresses which
would otherwise result from the wide temperature range to which this
area is subjected, approximately 4 K minimum to 650 K maximum. Thermal shields of gold-plated foil are used to prevent the temperature
rising further, which would result in the need to use heavier structural materials. The reaction chamber/thrust structure assembly for
both stages can be moved through $\pm$ 1° for steering purposes.

Suitable structural materials have been identified following a
selection process which considered the environment, in particular
thermal covering both elevated and cryogenic temperatures, materials
abundance, magnetic properties, nuclear radiation effects, fabrication
and electrical properties. From this selection procedure molybdenum
was chosen for the reaction chambers, with aluminium, titanium and
nickel alloys being selected for other parts of the structure.

2.6 Erosion Protection Systems

One problem which has to be considered for interstellar probes is that of erosion due to particle impact. This problem becomes more severe the faster a probe travels. Daedalus has a coast speed of about 12% of the speed of light, and an analysis has been performed on the effects of particle impact as a result of that speed. Two cases have been considered, that of interstellar coast and that of Barnard's Star system encounter, the latter being the worst case due to interplanetary debris.

In the interstellar coast phase dust grain impact is the most severe form of bombardment. For collision with typical grains with a mass of 10^{-16} kg, the energy transfer between the nuclei of the vehicle and nuclei of the particles would lead to permanent damage to the vehicle, VIA nuclear disintegration, vaporisation, displacement, etc. An erosion shield is therefore required to protect the vehicle. Based on calculations of the shield mass loss due to permanent changes due to ablation heating, it was found that a 7 mm thick beryllium shield with a diameter sufficient to protect the second stage engine (55 metres diameter) would offer adequate protection during the interstellar coast.

During the encounter phase of the mission it is expected that there will be an increase in the probability of a collision with larger particles. Based on debris distribution estimates for the Solar System it was determined that the vehicle protection system for encounter must be capable of handling debris up to a mass of 500 kg. It was proposed to deploy a cloud of particles approximately 200 km ahead of the vehicle to act as a screen to dispose of large particles in the flight path. The screen particles required are very small, and it was estimated that a cloud with a total mass of 6 kg was sufficient, and that it would need replenishment every 14 hours during encounter in order to maintain its density. This cloud will be deployed either by small craft, or a projector device in the front of the vehicle.

The small size dust and debris which is left after impact with the cloud will be stopped by the main vehicle shield, and an equal thickness to that used during the cruise phase would be employed for this purpose.

2.7 Collection of Helium 3 from the Jovian Atmosphere

Daedalus requires 30,000 tonnes of Helium 3 and 20,000 tonnes of deuterium as thermonuclear propellants for each vehicle. Helium 3 is a very rare isotope, with a natural abundance of about one part in 100,000, and on Earth is produced by the neutron bombardment of Lithium 6. To do this for Daedalus, manufacturing the propellant supply over a 20 year period, would demand reactors with a total power level of 200 million megawatts.

However, 17% of the atmosphere of Jupiter is helium, and assuming the abundance of Helium 3 relative to Helium 4 is the same there as upon Earth there is still enough of the isotope to fuel millions of Daedalus vehicles if we had the patience. Because Jupiter lacks a defined solid surface, we would float separation plants within the atmosphere beneath hot-air balloons, using the waste heat from the power reactor to generate lift. The propellant collection rates appear modest - a total rate of 47.5 gm/sec of Helium 3, and 31.7 gm/sec of Deuterium over a 20 year period - but in the process some 28 tonnes of Jovian atmosphere must flow through our separation plants each second. With these flow-rates heat must be transferred exactly between inlet and outlet gases. A $1^{\circ}K$ temperature difference between inlet and outlet represents an energy leak of 400 MW.

With these quantities it is expected that many separation plants would be floated within the Jovian atmosphere, both to bring the size of each down to manageable proportions, and to allow for natural losses. These would be serviced by a regular orbital "shuttle".

The deep gravity well of Jupiter at first sight seems to be a formidable obstacle to lifting propellant from the atmosphere into orbit. The orbital velocity is no less than 42.1 km/sec. However, the natural rotation of Jupiter means that a point on the equator already has a velocity of 12.7 km/sec, and in addition it should be possible to fly ramjet vehicles through the largely hydrogen atmosphere at speeds of up to 10 km/sec. The propulsion requirements should be within the capability of nuclear rockets available at the time.

Collecting Helium 3 from the atmosphere of Jupiter means a big reduction in power required, from 200 million megawatts down to about 500 megawatts.

2.8 Computers and On-board Maintenance Systems

In view of the lengthy unmanned flight time of the vehicle, the lack of access to comprehensive supporting repair facilities and the need for a high probability of success, the design included provision for advanced computing facilities. These were capable of directing the sequencing and control of all experiments, probes and sub-probes, data reduction and storage, navigation and attitude control, the normal operational control of all supporting systems and the detection, location and rectification of all faults, involving component repair on board.

The vehicle was designed to carry a hierarchy of computers with executive authority for the control of all experiments and supporting systems. The computer software, also in hierarchical form, must be capable of varying the operational goals embedded within itself in order to exhibit adaptive behaviour. Not only must the computers control full on-board repair facilities but must also themselves be de-

signed for repair while continuing to operate, i.e. they must be
capable of fault-tolerant fail-safe operation.

The aim of the mission could only be met if on-board facilities
were carried for extensive repair-by-repair techniques. Where possible
sub-systems will be designed to repair themselves, but the majority
of faults still require external intervention for rectification. The
repair facilities, and the spares and materials needed to feed them,
were split into two parts - one static part for off-line repairs and
a second mobile part which either repaired faulty items in situ, or
removed them for feed-in to the static part. This role was assigned
to independently-mobile robot repair vehicles, called "wardens", and
a minimum of two, nuclear powered and weighing 5 tonnes each, were
considered necessary to permit co-ordinated action and mutual repair,
where necessary.

3. FAST STARSHIPS

3.1 Introduction

The vehicle design discussed in Section 2 was concerned with a
concept representing more or less an extension of the current theme
of planetary exploration with remote probes. It is certain that
interstellar exploration, if carried out at all, will not end with
crude probes of the Daedalus type. Quite how it will develop is
impossible to judge, but two different general approaches seem to
exist.

The first is to aim for more and more powerful engines so that
vehicles travel ever closer to the speed of light. As this velocity
is approached the effects predicted by the theory of relativity
become manifest, and time as given by clocks on the vehicle runs slow
compared with time on the Earth. If the vehicle is swift enough, the
crew could travel the accessible universe in a human lifetime, as
shown in Table 2. The people remaining on Earth would, of course,
be long dead by the end of the mission, and indeed the Earth and Solar
System may no longer exist if the mission was long enough (in terms
of both time and distance).

The second approach is that of missions which only achieve very
low velocities, of the order of 1% of the speed of light or less.
These missions have propulsion requirements which are well within the
technology discussed in project Daedalus. In this mode of travel,
extremely long trip times must be tolerated and we are forced to the
concept of multi-generation vehicles with crews dying after their
natural lifespan, only to be replaced by their children.

Table 2.   Interstellar Journey Times

Assumptions are acceleration of 1 Earth gravity (9.81 msec^{-2}) during the initial half of the journey, followed by deceleration at 1 g during the second half.

Journey	APPROXIMATE distance		Journey time	
			Earth-time	Ship-time
	(m)	(ℓyr)	(yr)	(yr)
Alpha-Centauri	4×10^{16}	4	6	3.5
Centre of our Galaxy	10^{20}	10^{4}	10^{4}	18
Nearest Spiral Galaxy (Andromeda)	10^{22}	10^{6}	10^{6}	27
Postulated limit of physical universe	10^{27}	10^{11}	10^{11}	49

In this section we will discuss the prospects for fast, relativistic starships. This concept is treated first, not because we believe it to be the most probable, but because it is usually assumed in discussions of interstellar travel without consideration of the difficulties inherent in the methods proposed to achieve it.

3.2   The Antimatter Rocket

The ultimate rocket propulsion system which appears possible is the photon rocket, where matter is completely annihilated and the resultant energy is radiated away to produce thrust. There is no possibility, within our present-day knowledge of physics, of a vehicle containing its own energy source and propellant being better than this system.

If a direct beam of electromagnetic energy could be used to provide thrust, the vehicle would have an exhaust velocity equal to the speed of light. It would, however, consume power at the rate of 300 MW per Newton thrust, and for a vehicle to accelerate at 1 g it would need to produce power at the rate of 3,000 MW per kgm of vehicle mass.

In order to achieve sensible mass ratios, it would be necessary to convert mass very efficiently into energy. The presently known nuclear processes are only able to liberate less than 1% of the available mass as energy, and hence it is often proposed that matter-anti-matter reactions, such as the proton-antiproton reaction (Fig. 3), should provide the energy.

Antimatter, however, is totally absent from our part of the universe as far as we can tell, and hence has to be manufactured. In the process an amount of energy equal to double the mass of anti-matter produced is consumed, and hence it is astronomically expensive. The cost of production and energies required make onboard production

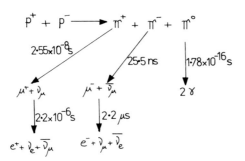

Figure 3. Proton-antiproton annihilation reaction (note that if, as seems likely, the electron and positron annihilate, the result of the reaction is simply neutrinos and gamma rays).

completely unfeasible. However, even assuming production prior to the trip, another problem is that of storage. Solid walls would, of course, result in the annihilation process occurring. Storage as a high temperature plasma in a magnetic field, or as a magnetically levitated antimetal, may be possible answers. In the first case the density is impractically low, giving a very heavy vehicle, and in the second case the nuclear technology to manufacture, say, anti-lithium is very difficult.

A further problem is that of radiating the energy away from the vehicle with a very high efficiency, as the deposition of even a small amount of the quantities with which we are concerned would result in the rapid vaporisation of the vehicle. Reflectivities of at least $1-10^{-6} = 0.999999$ are required. Silvered mirrors have only $1-10^{-2}$ in the infra-red range, and even less in the UV. It has been suggested that the containment of hot plasmas by magnetic fields, and the total reflections of energetic photons by a very dense electron gas may offer some promise.

It may be added here that if it is not possible to achieve 100% mass conversion, it is more efficient to use the energy obtained from what is converted to accelerate the remaining mass, rather than to radiate the energy in a beam and simply jettison the extra mass. Thus if the conversion was 50% efficient, relativistic mechanics tell us that if the remaining 50% of the mass was accelerated using this energy, then an effective exhaust velocity of 86.6% of the speed of light would be obtained. If the unconverted mass was dumped, only 50% of the speed of light would be attained.

3.3 The Interstellar Ramjet

Another contender for a vehicle capable of relativistic speeds is the Interstellar Ramjet (ISR) conceived as a way to avoid the staggering vehicle masses demanded by a craft which carries its own fuel and propellant, and which attempts to reach high velocities. In this concept all the propulsion system consumables come from outside the vehicle, and so its range and speed are in principle unlimited. The system has, apparently, enormous potential, but its practical difficulties are almost endless.

In concept the ISR is very simple. Interstellar space is filled with hydrogen gas in varying quantities, some in the neutral and some in the ionised form. This gas is collected by an intake on the interstellar vehicle and is fed into a thermonuclear reactor which converts a certain amount of the mass into energy in a fusion reaction. This energy is then used to accelerate the unconverted matter out of the rear of the vehicle to provide a thrust force.

The particle intake on the vehicle is responsible for most of the problems which arise when the ISR is studied in detail. It can be shown that the ratio of the frontal area loading density of the ramjet to the charged particle number density must have a value at least as large as 10^{-13} kgm/m^2 per unit reactive nucleon number density to attain an acceleration of 1 g_0. This implies that, for a vehicle mass of 1,000 tonnes, operation at highly relativistic velocities in a region of high hydrogen density (10^9 particles/m^3) requires an intake radius of about 60 km, and operation in a low density region (10^6 particles/m^3) needs a radius of 2,000 km. These densities are the approximate limits of hydrogen ion clouds in interstellar space.

Such large sizes effectively rule out the use of solid materials, as the structural densities would be far too large. In addition, the surface of such intakes would be continually heated and eroded by the bombardment of high energy particles which are being collected, and would rapidly fracture and lose their structural integrity.

Therefore the intake is usually considered to be a magnetic type, in which hydrogen ions are collected by the magnetic field lines and fed into the reactor mouth at the front of the vehicle. If the hydrogen in space is not ionised this can be easily accomplished by shining a light beam ahead of the ISR, thus causing photoionisation to occur. A schematic picture of such an intake is shown in Fig. 4 (not to scale). The spatial limit of the intake is given when the magnetic flux density, B, is equal to that of the interstellar magnetic field ($B_0 = 10^{-10}$ tesla). Inside the intake B is greater than B_0, and rises to a maximum value of B_m near the mouth of the reactor.

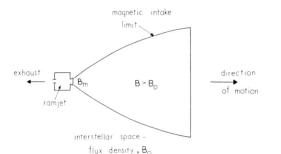

Figure 4. Ramjet magnetic intake configuration.

The initial theory of the ISR has been extended to consider the
limitations placed upon the vehicle motion by the vehicle structure.
It is assumed that the section of the vehicle which carries the sources
of the magnetic fields must be strong enough to withstand the forces
exerted upon these sources by the fields which they create. That is,
it is assumed that the magnetic field energy must be balanced by mechanical forces.

Known materials are limited in the values of their maximum tensile strengths, i.e., the amount of force that they can withstand before breaking up. Now, the magnetic flux density of the intake increases, in order to be able to confine the charged particles, as the velocity of the ISR increases; and so the energy which must be contained by
the structure also increases.

Therefore there must come a point where the acceleration of the
vehicle must be reduced to avoid the breakdown of the vehicle structure
by the magnetic forces. An expression for this cut-off velocity can
be derived, showing that the cut-off is proportional to the ratio of
the maximum tensile strength to the density of the material used.
Acceleration is still possible beyond this cut-off but the level must
be reduced to compensate for the added strain produced by the magnetic
field with the increasing velocity.

The effect of the decrease in the acceleration is shown in Fig.5,
where the distance travelled by the ISR is shown as a function of the
ship time. It can be seen that the acceleration rapidly decreases to
a point where no large increase in distance occurs over a period of
time.

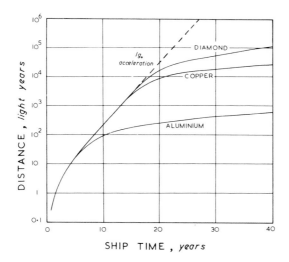

Figure 5. Performance curve for a proton ramjet
in a low density interstellar region.

    The problem of the confinement of ionised material by the use of
a magnetic field which gradually increases towards a maximum at the
vehicle is similar in many ways to that of particle confinement in
fusion reactors which make use of a 'magnetic mirror' geometry, and
indeed much of the physics is found to be equivalent. The obvious
difference is that the mirror seeks to achieve complete confinement
with zero particle transfer past the point of maximum field strength,
while the intake attempts to compress a large volume of material into
a dimension of the order of the reactor mouth without also causing a
large particle reflection fraction.

    This situation can be analysed as follows. First, all particles
with more than a certain amount of their initial momentum in the component transverse to the direction of motion of the vehicle are reflected by the magnetic field maximum, i.e., they are mirrored. Second,
all particles which have an initial radius of gyration larger than a
certain value are not injected into the vehicle powerplant, even if
the particles are not reflected, i.e., they are not confined. These
calculations allow values of the intake fractions of the ramjet to be
estimated.

    In order to allow investigation of the fundamental physical limits
of the intake a constraint-free technology, which places no limit upon
attainable flux densities or structural strengths, is considered.
Assuming that a proton fusion reactor will become feasible (which is,
of course, rather a large assumption in itself) allows a numerical
value of the probable necessary particle collection rate to be derived,
and this value can be used in conjunction with the intake fractions to
determine the minimum possible intake dimensions.

    The minimum values of the intake radius, for operation in a high
density region, are shown in Fig. 6. The scale used does not show the
behaviour of the radius near $v/c = 1$ clearly. Above about $v/c = 0.9$
the radius decreases rapidly. It can be shown that the ISR requires
an initial velocity component, and so the lowest velocity considered
here has been set equal to $10^{-4}c(v = 30$ km/sec$)$. The values in Fig.6
can be compared with those required by the frontal loading density
criterion for 1 g_0 acceleration, and it can be seen that the values
given by the present calculation are much larger.

    More serious, however, is the problem of the magnetic flux den-
sities which are needed to confine the particles to the reactor. Modern technology is only able to produce superconducting solenoids with
critical flux densities of 100 tesla and values of 1,000 tesla may
be feasible. Introduction of this technology constraint into the
calculation increases the required intake radius, as shown in Fig. 6.

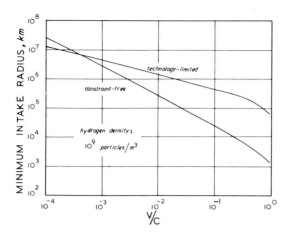

Figure 6. Minimum
intake radii in a
high-density region

3.4 Catalytic Ramjets

A further major problem with the concept of the interstellar
ramjet proposed above is that the proton burning cycle is too slow.
The chain can be represented by

$$p + p \longrightarrow {}^2D + e^+ + \nu$$

$$^2D + p \longrightarrow {}^3He + \gamma$$

$$^3He + {}^3He \longrightarrow {}^4He + 2p$$

The net result of the cycle is the conversion of four protons into
a ^{4}He nucleus with the release of 19.53 MeV of potentially usable
energy plus 0.263 MeV in neutrino losses. However, the weak interaction
represented by the first equation above has a very small cross-section
at essentially any temperature, and its applicability for the production of propulsive power is questionable.

The deuterium reactions

$$^2D + {}^2D \longrightarrow {}^3He + n$$
$$\searrow T + p$$

have cross-sections that are over 20 orders of magnitude larger than
the p + p reaction, but the energy yield is smaller and there is
only about one deuteron for every 10,000-100,000 protons in interstellar space.

A possible method of using the interstellar hydrogen as fuel, but
avoiding the slow p + p reaction rate has been suggested, involving

using a catalytic nuclear reactor chain, where one element in the cycle acts only as a catalyst and is returned to the plasma after completion of the cycle. The catalyst fuel is therefore not depleted, and can therefore be carried along in the vehicle. The two proton burning catalytic chains suggested are the CNO Bi-cycle and the Ne-Na chain, illustrated in Fig. 7. In these cycles the slowest reaction rates are $\sim 10^{18}$–$\sim 10^{19}$ times greater than the p + p rate.

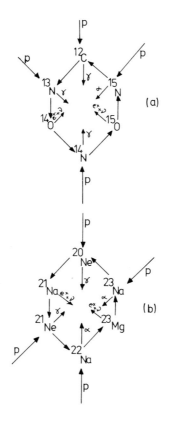

Figure 7.   (a)   Hot CNO bi-cicle, catalyst ^{12}C.
                 (b)   Ne-Na cycle, catalyst ^{20}Ne.

Unfortunately, the use of these chains requires a fusion reactor which involves levels of technological extrapolation comparable to those discussed in the previous section. In particular, the plasma confining magnetic field in the reactor is in the range of 10^3–10^4 tesla with an ion temperature of $\sim$ 100 keV. The basic stability of such a system is uncertain, let alone the technological difficulties associated with producing such enormous magnetic fields.

3.5   Ram-Augmented Interstellar Rockest (RAIRs)

A possible way of alleviating the problems associated with reactor

physics technology has been suggested – that of the Ram Augmented
Interstellar Rocket (RAIR). This vehicle (Fig. 8) carries a nuclear
fuel supply and exhausts the reaction products to produce thrust.
However, it enhances its performance by scooping up the interstellar
medium and using its momentum and kinetic energy to augment that of
the rocket. It thus still has the intake problems associated with the
ISR, but now the reactor problems fall into the range of reasonable
extrapolation of present technology, or at the most not too far beyond.

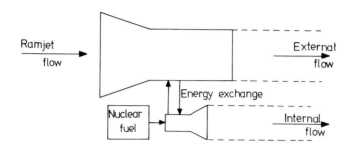

Figure 8. Flow
diagram of the
ram augmented
interstellar
rocket

By choosing the relationships between the vehicle velocity, the
exhaust velocity of the interstellar medium processed by the vehicle,
and the exhaust velocity of the reactor products it is possible to
optimise the overall vehicle performance. In an ideal case all of
the installed energy in the vehicle would be imparted into the final
kinetic energy of the vehicle. A practical machine would achieve
rather less than this, but the improvement over the straightforward
rocket will still be considerable.

There are three possible phases of operation in the general case
of an inefficient RAIR. At first, the jet velocities are greater than
the vehicle velocities. During this phase of operation, energy is
being transferred from the fusion reaction products to the external
flow. This reduces the internal flow exhaust velocity, and increases
the external jet velocity. Secondly, the ramjet flow gradually falls
to zero during the first phase, and during the second phase the vehicle behaves as a pure fusion rocket with all the fusion energy being
carried by the reaction products. Thirdly, the external mass flow
begins to increase again, but now the jet velocities are less than
the vehicle velocity. Energy is now being taken from the external
flow and added to the internal flow, producing a rocket exhaust which
has a higher velocity than the straightforward fusion rocket. Indeed,
it is interesting to note that at very high velocities, the nuclear
energy contribution to the jet velocity is so small that the mass
being ejected from the vehicle can be inert as far as energy production is concerned, all the energy necessary for propulsion being derived from the kinetic braking of the external flow.

3.6 Problems of Interstellar Erosion

The regions of interstellar space are not devoid of matter, but
contain particles of various kinds and with varying densities. Any
vehicle which attempts to cross these regions will be bombarded by
this matter, and the effect of the erosion caused by this will be
severe at high velocities.

For convenience the effects of the bombardment of interstellar
material can be summarised under two general headings:
  a. Processes which result in the heating of the material, but
  which do not result in permanent changes in the material. This
  will be caused by the deposition of energy (both direct kinetic
  energy and via the absorption of radiation) from protons and
  electrons.
  b. Processes which result in permanent damage to the material,
  including mass loss by ablative heating. This will be caused
  by the bombardment of dust grains at velocities well below the
  relativistic regime, and will be dominant for slow-speed ships.
Some values for the mass loss for a vehicle with an ablating erosion
shield are given in Table 3. Here the material used as an example is
beryllium. As relativistic speeds are approached the thickness of
material eroded approaches, and exceeds, values of a metre of material
per square metre per year. For even "small" vehicles with reasonable
trip times, this represents a staggering mass loss.

Table 3. Values of mass loss rate and thickness of material eroded
          from a beryllium erosion shield, as a function of vehicle
          velocity (average interstellar grain density taken as
          $\sim 10^{-23}$ kgm m^{-3}).

v/c	Mass loss rate (tonnes m^{-2} yr^{-1})	Thickness eroded (m m^{-2} yr^{-1})
0.5	2.14×10^{-2}	1.16×10^{-2}
0.8	0.21	0.12
0.9	0.64	0.35
0.99	10.3	5.5
0.999	115	62

4. WORLD SHIPS

4.1 Introduction

The use of fast starships for interstellar flight has received
a good deal of attention, but an alternative which has been unduly
neglected, except in broad concept, is the use of much slower vehicles
having coast velocities of 0.01%-1.0% of the speed of light.

In concept, such a vehicle would leave its home system carrying
a community whose objective was simply to live a normal existence,
with the slight variation that instead of following the orbit of
the parent planet they would follow a highly eccentric orbit through
the Galaxy, such that future generations would have the option of
exploring or colonising other planets as they were encountered. To
compensate for the resulting long journey times the vehicles must
increase in size to such an extent that they should be considered to
be small worlds in their own right. Some general proposals for world
ships are listed in Table 4.

Whilst the underlying motive of such missions must obviously be
to spread through the Galaxy this could not possibly be the immediate
goal for the occupants, since the achievement of such a task would lie
many generations in the future.

Table 4.   Some proposed world ships.

Originator	Vehicle Mass (tonnes)	Population	Flight Duration (years)
R H Goddard (1918)	?	?	10,000– 1,000,000
J D Bernal (1929)	?(10 mile sphere)	20,000– 30,000	∿ millenia
J G Strong (1965)	10^8	100–150	∿ 100
F J Dyson (1968)	4×10^5	200	∿ 100
F J Dyson (1968)	4×10^7	20,000	∿ 1000
R D Enzmann (1973)	1.2×10^7	2,000	∿ 100
G L Matloff(O'Neill)(1976)	6.5×10^5	10,000	∿ 1000
G L Matloff(O'Neill)(1977)	2×10^4	100	∿ 200
M G de San (1978)	2×10^{10}	1,000,000	∿ millenia

We need not concern ourselves too deeply here with motives, since
we are only interested in plausible concepts of interstellar transport,
and world ships seem to satisfy that requirement, for whatever reason
they may be built.

The constraints on world ships appear to be twofold. First,
materials considerations suggest a size limit to the vehicles. Second,
the fact that the ships would be so massive limits the velocity to
which they can be accelerated, from an energy availability point of
view.

4.2 World Ship Size

A vehicle which would be in operation for perhaps many millenia
must be capable of providing a stable environment, presumably not too
dissimilar to that of the planet of origin. It appears a reasonable
assumption that the inhabitants of the starfaring planet would wish to
take their culture and basic environment with them and that to do this

they will fabricate the largest ships practicable and then use them
in fleets.

We have argued elsewhere that there are probably a considerable
number of planets in our Galaxy which are close to the Earth in environment. Whilst we do not claim that they will evolve life to our
level of complexity, it does at least provide the correct starting
conditions and we therefore feel justified that some world ships, if
they are built, must be designed to provide Earthlike environments.
In order to assess, therefore, the characteristics of a world ship
we adopt its basic environment as shown in Table 5. (The following
discussion of world ship structure, size, propulsion, etc., is based
upon preliminary, unpublished work by the authors.)

Table 5. World ship environment specification

Acceleration field	10 m sec^{-2}
Atmospheric pressure (O_2, N_2)	10 N m^{-2}
Ambient temperature	$290^\circ K$
Cosmic ray background dose rate	< 100 mrem/year
Interior regolith density	$2,500 \text{ kgm m}^{-2}$
Interior regolith depth	5 m
Illumination	10^5 Lux, daily and seasonally varied and corrected for latitude desired.

The simplest structure capable of providing these conditions is
a rotating cylinder with the occupants on the interior. Since the
mechanical stresses involved limit the dimensions which can be achieved,
it is proposed that maraging steel, which has very good fracture tough-
ness characteristics and hence is tolerant to manufacturing defects,
would be used for the outside shell. The materials required (iron,
nickel, cobalt, titanium, aluminum) are all available in large quantities in the asteroid belt.

For a world ship we would expect very steady structural loads,
since the object is to provide a stable environment. Analysis of
stresses and loads on a vehicle constructed from the above material
allows us to calculate the wall thickness as a function of ship radius,
and this is shown in Fig. 9. It can be seen that the wall thickness
increases rapidly for a radius greater than $\sim$ 8-9 km (which coince-
dentally was about the size of A C Clarke's RAMA). With this radius
the wall would be $\sim$ 4 metres thick.

To determine the length of the cylinder, and hence the mass of
the ship, we need to estimate the population required. This is at
present subjective, and previous estimates (Table 4) have varied between a few hundred and a million. However, considering the situation
on Earth at present it appears that towns can be self-sufficient at
between 100,000 and 200,000 inhabitants. Bearing in mind the length

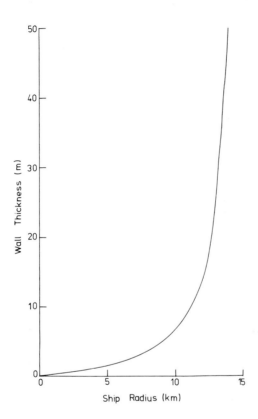

Figure 9. Wall thickness as a function of
ship radius for an earth-like environment

of the journey and the requirement for long-term stability, we feel
this is a reasonable minimum figure. If we assume a working population of 100,000, and a 40% occupation-related factor, then we require
a total number of inhabitants of 250,000.

Population densities on Earth vary considerably, as seen in
Table 6. We assume a mid range value of 100 per square kilometre,
giving a cylindrical surface area of the ship of 2500 km^2, corresponding to a length of 50 km. Much higher population densities exist in
cities and have been proposed in space colony studies, but we do not
consider such large densities to be conducive to long-term stability.
(The length of the world ship is, of course, directly proportional to
population density.)

Table 6. Population densities on Earth (c.1980)

Country	Population Density (km^{-2})
Netherlands	319
Japan	281
United Kingdom	241
France	90
USA	22
USSR	11
Australia	2

With hemispherical end closures, the total mass of the world ship
structure, the regolith and the atmosphere (obtained either from the
home planet or a gas-giant planet) is estimated as some 140 gigatonnes
(1.4×10^{11} tonnes). While this picture of a world ship is obviously
heavily biased towards the anthropocentric viewpoint, it does at
least give an indication of the scale of engineering which must be
considered.

We have discussed what we believe to be close to the limits of
practical scale for a world ship with an acceptable Earth-like environment. However, by using flotillas of such vessels very large communi-
ties could be in transit between stars. For example, 100 vessels
could carry 25 million inhabitants, or the typical population of a
European nation. Smaller transit vehicles could shuttle between the
world ship, in a manner similar to present-day aircraft travel between
countries. The total production capabilities of such a fleet, bearing
in mind the technological level implied by its mere existence, should
be adequate to colonise and modify new worlds as encountered.

4.3 World Ship Propulsion

In order to propel a vehicle the size of a world ship the engine
power would need to be about 10^{18}-10^{19} watts, or about a million times
the entire artificial power output of the Earth. Such a power level

would be adequate to achieve a cruise velocity of about 0.1% of the speed of light after a boost period of about 1 to 10 years. Since a world ship must be able to command such engine power at any time, even when light years and millenia away from home, the power source must be on-board. The only practical mechanism currently known is the external nuclear pulse rocket. We may follow the calculation through for the 140 gigatonne vehicle discussed above. It can be shown that for nuclear vehicles propelled to relatively low velocities such as those here, the most economic fuelling occurs with a mass ratio (fully fuelled mass/empty mass) of 4.9, based on the assumption that nuclear fuel is expensive and inert propellant (assumed to be hydrogen) is essentially zero cost. Assuming the tankage to be expendable and part of the propellant, we therefore have a vehicle departure mass of about 700 gigatonnes of which some 550 gigatonnes is nuclear fuel and propellant. Assuming an initially installed total velocity gain of 1% of the speed of light (that is, a maximum velocity of 0.5% C) then the exhaust velocity of the engine would be $\sim 2 \times 10^6$ m/s. Assuming further that the nuclear fuel is deuterium extracted from the atmosphere of a gas-giant planet, and that the burn-up obtained in the explosions is 15%, we find that the ratio of propellant mass/fuel mass is about 30, i.e., about 3% of the propellant carried would be nuclear fuel. Thus, such a world ship would carry pulse charges containing about 20 gigatonnes of deuterium, a quantity readily available in the atmospheres of gas-giant planets.

Assuming an initial acceleration of 10^{-2} m/s^2 (1 milligee) the vehicle thrust would be about 7×10^{12} Newtons with an engine flow rate of about 4×10^6 kg/sec. If this consisted of nuclear pulses at 1 sec intervals it would represent the detonation of 130 tonnes of deuterium in each pulse with an explosive yield of about 2 gigatonnes of TNT equivalent. Such explosions have been discussed as feasible. It is assumed that the pulse units would have internal trigger mechanisms in the interest of reliability, since it would not then be necessary to provide a device capable of many repetitions as in Project Daedalus. The world ship above might carry some 140 million pulse units in total.

If we assume the propellant storage at 13°K and a reasonable allowance for packing, the propellant volume would be about 9×10^{12} m^3, or a cylinder the same diameter as the world ship and 40 km in length. Thus, including the propulsion thrust transfer mechanism and its shock absorbers, the world ship would be about 16 km in diameter and about 100 km in length.

The engine mechanism would probably consist of a simple massive pusher plate mounted via shock absorbers to the ship. It would be protected at each pulse by a tiny fraction of the hydrogen propellant. The whole mechanism would be designed to receive the momentum of the detonation and delay and smooth the load profile so as to accelerate the ship smoothly. Such mechanisms were studied in detail in Project Orion (1958-1965), the most serious nuclear pulse vehicle activity which has been carried out to date. Although the nuclear explosions would produce neutrons, the hull of the world ship has more than adequate thickness to attenuate these to a harmless level.

The fuel and propellant are the most abundant in the Universe. Given a stop in any planetary system it would be possible to refuel, and assuming adequate on-board manufacturing capability, new tankage and pulse triggers could be manufactured in the new system ready for departure, if necessary.

The above description represents an indication of the scale of engineering which we believe is possible. The only limit which we can see is the energy resources available from the atmospheres of gas-giant type planets, and these appear to be vast, based upon reasonable model of planetary system formation.

5. CONCLUSIONS

With our present level of science and technology we are not able to construct starships. However, several concepts exist whereby small, relatively high speed (> 10% c) probes and very large, slow (< 1% c) vehicles may reasonably be discussed. The technology to build very fast (> 90% c), small vehicles should also eventually emerge. In addition, the results of the Project Daedalus study indicated that the building of a first, relatively crude, interstellar probe would be possible with only modest extrapolations of present-day science and technology.

We are not justified, therefore, in assuming that such vehicles will never be possible - indeed, we would argue to the contrary. One may dispute the motivations behind such programmes, but its practicality cannot form a basis for dismissal.

We can therefore reasonably turn to the more general conclusions, relevant to the question of how advanced civilisations will manifest themselves, and ask questions regarding the nature and appearance of such vehicles, assuming that they are found to be both possible and practical.

Small probes may remain relatively obscure, having limited energy release and interaction with the target. This may not, of course, be the case for a probe which decelerated at its target and entered an extended exploration mode.

The small, fast vehicles and the large, slow vehicles would, however, be very energetic objects, releasing powers of greater than 10^{18} watts on average. In the case of the large, slow vehicles this release would probably be in the form of regularly repeated pulses of brief duration with energies greater than 10^{18} joules. The energy will be emitted from the vehicles with an efficiency almost identically

equal to unity in order for the vehicles not to vaporise. The interaction of the exhaust plume of a fast ship with the galactic magnetic
field should be an observable phenomenon, involving ions in the 0.1-10
MeV range and scales of 1-10 light years in length. The exhausts of
slow ships will produce plumes of highly ionised matter, together with
neutrinos, gamma rays and other particles.

Both types of ship will destroy any matter which lies ahead of
them, either by injection into on-board power plants, or removal from
their line of sight. The proper motion of both types of source as
viewed from the Earth would be large, and would be especially apparent
in the case of relativistic vehicles.

Such processes as those above should cause the radiation of energy at very high levels across the entire spectrum from radio wavelengths, through infra-red, visible, ultra-violet, into X-rays and
γ-ray spectra. At the power levels we consider necessary, these are
the sorts of vehicle that should be instantly noticeable, even when
compared to the scale of energy releases found in the rest of the
Universe.

While it is not our intention to discuss motivation in the present paper, it is relevant to consider the possible numbers of vehicles
that a civilisation could launch. This can be examined in terms of
several limiting factors:
  a. quantities of structural materials available
  b. quantities of propellants available
  c. quantities of nuclear fuels available

The quantities of propellants for rocket vehicles are limited
to the availability of hydrogen and helium, and is about 2×10^{24}
tonnes within the Solar System. The deuterium resource is about
3×10^{20} tonnes. The availability of maraging steel from the asteroids is no more than $\sim 10^{17}$ tonnes.

Thus, even assuming only 0.1% mining of deuterium, structural
materials would appear to be the limiting factor. Even so, this
would be sufficient for about half a million world ships. It is
difficult to assess the material limits of fast starships, for they
may be dominated by the availability of materials other than steel,
such as molybdenum and niobium, which would either be mined from
natural resources or bred. Even so, the numbers would probably be of
the order of the above estimate.

The vehicles SHOULD be there and we SHOULD see them, but we DON'T.
This apparent paradox leads us back to some of the central themes of
this present volume, and possible resolutions of the problem are more
fully discussed elsewhere. In concluding, we would list the three
most obvious and reasonable alternatives, i.e.,
  a. We do see signs of vehicles in passage among the stars
      but we are too naive to recognise this.

    b.   There are no civilisations at present in our Galaxy
        which are at a technological level more advanced than
        our own.
    c.   There do exist such civilisation but they do not develop
        and undertake interstellar voyages.

We believe that alternative (c) is not a realistic argument, but at
present we hesitate to choose between the other two explanations.
Only more work and the passage of time will answer our questions.

RADIO LEAKAGE AND EAVESDROPPING

Woodruff T. Sullivan, III
Department of Astronomy, University of Washington

Abstract: In addition to searches for purposeful signals, those attempting interstellar communication should also consider the possibility of
eavesdropping on radio emissions inadvertently "leaking" from other
technical civilizations. To understand better the information which
might be derivable from radio leakage, the case of planet Earth is considered. The most detectable and useful escaping signals arise in a
few ultra-powerful military radar systems and in normal television broadcasting. A model including over 2,000 television transmitters is used
to demonstrate the wealth of astronomical and cultural information
available from a distant observer's careful monitoring of frequency and
intensity variations in individual video carriers (program material is
not taken to be detectable). Observations of the Earth's leakage radiation, observed at Arecibo as reflected from the Moon, are also presented. It is concluded that, given our present modest understanding of
the cultural and technical evolution of civilizations, any initial interstellar radio contact has a priori even chances of placing us in the
role of eavesdropper or of intended recipient.

One can make either of two basic assumptions about our first contact with an extraterrestrial civilization: (i) that it will arise
through a purposeful attempt, perhaps through the use of an interstellar radio beacon, or (ii) that a civilization will be detected through
no special efforts of its own. The latter hypothesis, often called
eavesdropping, is concerned with the extent to which a civilization can
be unknowingly detected through the by-products of its daily activities.
While much thought has gone into the idea of purposeful contact, eavesdropping has been somewhat neglected; we will argue that it deserves
more attention.

The overall likelihood of contact through eavesdropping depends on
the nature and intensity of the civilization's "leakage", as well as on
how long that leakage continues. Very general arguments (see Oliver
and Billingham (1973)) show that radio waves provide the most economical and reliable means of contact at interstellar distances. This is
true not only for intentional contact, but probably for eavesdropping

227

M. D. Papagiannis (ed.), Strategies for the Search for Life in the Universe, 227–239.
Copyright © 1980 by D. Reidel Publishing Company.

as well. In any case, there can be no argument with the fact, first
discussed in print by Webb (1963) and by Shklovskii (1966), that the
presence of humans can already be detected at interstellar distances as
a result of the complex communications and transportation network spread
over our globe. Of course, we do not know how applicable our present
situation is to the more general case of all galactic civilizations
over all time. It may be that our present "leaky" state will soon be
terminated by advancing technology, but on the other hand it may continue for a very long time, perhaps even longer than any period in
which we might have the perseverance to send out purposeful messages.
If we are at all typical, then we should perhaps be also looking for
unintentional signals from others at least as diligently as for intentional ones. In order better to understand the principles involved,
let us now examine the appearance of the Earth from interstellar distances.

LEAKING RADIATION

     Detailed consideration of all parts of the electromagnetic spec-
trum reveals that it is radio waves which are by far the most important "leakage" from the Earth. For instance, nothing that we do with
visible light, not even exploding a hydrogen bomb, compares in the least
with the Sun's output. But at wavelengths from 1 centimeter to 30 kilometers our society has organized a host of activities on Earth which
give our planet an unnatural "radio signature": television and radio
broadcasting, radars used for weather, navigational and military purposes, "shortwave" communications ("hams", Citizens Band, taxis, police),
satellite communications, etc., etc.

     We now want to put ourselves in the "shoes" of an extraterrestrial
radio astronomer on a planet revolving about a star far from our Sun.
Which of these radio services would be "best" for our "eavesdropper" to
tune in on? Which is detectable to the greatest distances? Which potentially carries the most information of use to the eavesdropper? To
answer these questions one must study many factors including the power
of each service's transmitters, the frequencies and bandwidths involved,
types of antennas used, and the fraction of time spent transmitting.
(Readers desiring technical details on these and other matters should
consult Sullivan et al. (1978)). One example of these factors is the
general trade-off between the information content (TV picture, spoken
words, Morse code) of a transmitted signal and the distance to which it
can be detected. This can be understood by noting that one gets more
range by concentrating transmitted power at the fewest number of frequencies possible. But the information content of a signal is contained in the arrangement of its power among a number of neighboring frequencies and increases as we spread the power over a greater bandwidth.

     Three other important criteria in the evaluation of each radio
service are: (a) that the signal should be exactly the same from day to
day, (b) that the amount of sky illuminated by the transmitting antennas

should be large, and (c) that the number of transmitters on Earth should
be large. Regarding (b), remember that the radio waves from even a
stationary antenna can sweep out a large portion of the sky as a result
of the Earth's rotation. Furthermore, each antenna has a characteristic beam into which the transmitter power is directed. If an antenna
is designed so that the power is concentrated into a relatively small
region of the sky, the range of detection for the signal increases, but
at the expense of excluding many potential listeners.

ACQUISITION AND INFORMATION SIGNALS

Keeping the above factors in mind, an examination of all radio
services reveals two categories of strong signals escaping the Earth
that might be of interest to an extraterrestrial observer.

An acquisition signal merely announces our presence over a large
region of space by its very existence, but is not generally useful for
careful study because it fails to meet one or more of the criteria giv-
en above. An information signal, however, satisfies all three criteria.
At the present time on Earth some of the most important acquisition
signals originate from a half-dozen or so U.S. military radars (and
their presumed Soviet counterparts). These Ballistic Missile Early
Warning System (BMEWS) radars sweep out a large fraction of the local
horizon with extraordinarily powerful transmitters. The result is that
this "radio service" provides by far the most intense signals which
leak from our planet to a large fraction of the sky.

While BMEWS radars pass criterion (b) above, they fail (c) and
partially fail (a) because there are so few of these radars and they
often change their frequency of operation to avoid being jammed. Nevertheless, if an external observer used equipment comparable to the most
sensitive radio telescope on Earth (the 305-meter diameter dish at Arecibo, Puerto Rico), we calculate that a BMEWS-type radar could be detected as far away as 15 light-years. This distance includes only about 40 stars, but of course it is possible that our eavesdropper possesses
a much more sensitive radio telescope than we. If he had something
like the largest one ever proposed for Earth, namely the array of 1000
100-meter dishes called for by Project Cyclops (Oliver and Billingham
1973), he could detect a BMEWS-type radar at a distance of 250 light-
years. In this case at least 1,000,000 stars are possible candidates
for such an eavesdropper's location. But radio waves of course travel
at the finite speed of one light-year per year and thus it will take
until the 25th century, or 500 years from now, before all of these stars
have had a chance to be bathed in the radiation of our strongest military radars!

After picking up a BMEWS (or other) acquisition signal, the observer needs at least 100 times more sensitivity in his equipment to reach
the rich lode of information signals emanating from Earth. It turns
out that television broadcast antennas (or "stations") are the most

intense sources of such signals. All other services either have their
transmitter power spread over too broad a frequency band (for instance,
FM broadcasting and most radars) or they do not transmit continuously
(ham radio operators) or from the same location on Earth each day (ships,
aircraft). Many signals, such as medium-wave AM broadcasting and almost all shortwave communications, never even penetrate the ionosphere.
We thus concentrate on TV broadcasting - all other services which leak
from Earth are less intense and merely add to the background noise
which a distant observer would measure in the direction of our Sun as
seen in his sky.

But again note that TV broadcasting from Earth has been in existence for only 40 years. Figure 1 illustrates the phenomenal growth in

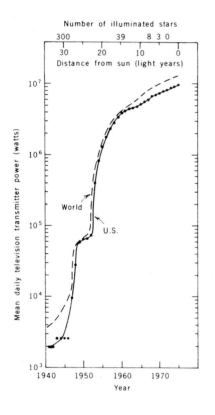

Figure 1. Estimated growth in time-
averaged transmitted power since television broadcasting began. The solid
curve for the United States is reasonably accurate since starting dates
for all stations were available, but
it was still necessary to employ a
model for the growth of transmitter
power and daily broadcast hours for
each station. The dashed curve for
the entire world is correct for 1975
and only estimated for earlier dates.
The increasing number of stars bathed
by the expanding power bubble (as of
1975) is indicated at the top of the
diagram.

intensity of the resultant ever-expanding "power bubble". On a cosmi-
cally infinitesimal time scale, the Earth has indeed become a very
bright planet, outshining the Sun by orders of magnitude in certain
narrow frequency ranges.

TV BROADCASTING SIGNALS AND ANTENNAS

In order to understand why television is so valuable to the eavesdropper as an information signal, we should discuss some of the characteristics of TV broadcasting signals. Perhaps the most important
facts are that there are a large number of very powerful TV stations on
Earth (Fig. 2), and that about one-half of a station's broadcast power
resides in an extremely narrow band of frequencies, only about 0.1 Hertz
wide, called the video carrier signal. The other half of the power contains the picture information and is spread out in a complex manner over
a far larger frequency range of about 5 MHz. Nowhere in this broader
region is the power per Hertz even a thousandth that at the video carrier
frequency. It would therefore be much more difficult (a factor of
2×10^4 is a good estimate) for the eavesdropper to receive full program material than to simply detect the presence of the carrier signal.
(Given the quality of most TV programs, we find this fact very reassur-
ing!) An observer near Barnard's Star, third closest to the Sun at a
distance of 6 light years from Earth, is thus about to receive television signals originating from the 1974 "Watergate" hearings, but he
probably cannot find out if Nixon was impeached! In the discussion below, we assume that only the video carrier signals of stations, not
program material, are detected.

The combination of reasonably high power and small bandwidth means
that the most powerful TV carrier signals can be detected (at optimum
frequencies of 500-600 MHz) from distances as large as one-tenth of
those discussed for the BMEWS radars. The narrow-band nature of the
signal also enables the observer to measure extremely accurate Doppler
shifts in the frequency of the carrier signal, allowing a determination
of the relative speed with which each station is moving to an accuracy
of about 0.0001 km/sec. Each station's signal thus contains information concerning the myriad motions in which its broadcast antenna par-
ticipates while anchored to the rotating and revolving Earth (see
Figure 3). Note that stations on a common channel will not fall pre-
cisely on top of each other's frequency because the combined effects of
engineering sloppiness, deliberate frequency offsets, and Doppler motions
all shift a station's video carrier frequency by much more than its
width (Figure 4). This means that our hypothetical observer could not
obtain a more favorable signal-to-noise ratio by trying simultaneously
to receive all the Channel 5's, for example, from a single region. The
problem of detecting radio leakage from the Earth as a whole is thus
seen to be essentially identical to the problem of detecting its single
strongest transmitter.

The beam patterns into which TV broadcast antennas radiate are
important to consider in such an analysis. It turns out that these
antennas (whose purpose, after all, is to broadcast to Earth and not to
the stars) confine the transmitter power to within a few degrees of the
horizon, but distribute it about equally in all compass directions.
Those radio waves directed above the horizon completely escape the
Earth's atmosphere, and even about half of those below the horizon

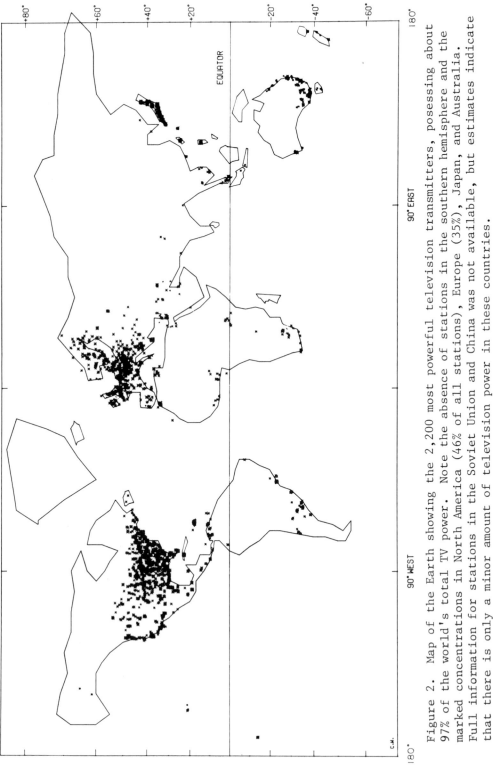

Figure 2. Map of the Earth showing the 2,200 most powerful television transmitters, possessing about 97% of the world's total TV power. Note the absence of stations in the southern hemisphere and the marked concentrations in North America (46% of all stations), Europe (35%), Japan, and Australia. Full information for stations in the Soviet Union and China was not available, but estimates indicate that there is only a minor amount of television power in these countries.

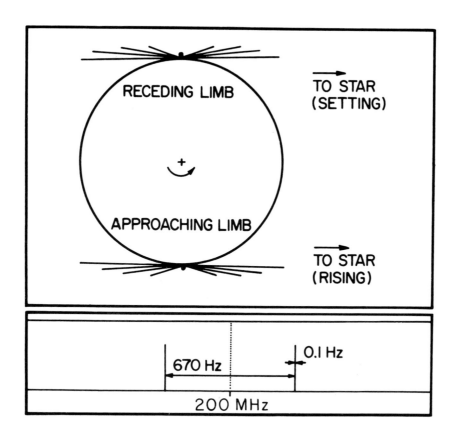

Figure 3. (Top) Sketch of two TV broadcasting antennas as seen from
above the Earth's pole. (The sketch also applies to a single station
as it would be seen at 12-hour intervals.) The length of a particular
line of radiation indicates the relative amount of power "beamed" in
that direction. As seen from a distant star located to the right, both
stations are at maximum intensity, but one is just coming into view and
the other is just disappearing. From the point of view of the stations,
one sees this star rising and the other sees it setting. (Bottom)
Radio spectrum of the two stations' video carriers as measured at the
distant star. Both stations are assumed to radiate from Earth with the
same rest frequency (dotted line); the observed frequencies are different as a result of the Doppler effect arising from the Earth's rotation.
The numbers given are for stations taken to radiate at 200 MHz (approximately U.S. channel 11) on the equator, and are typical of those for
most stations.

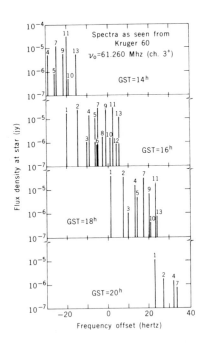

Figure 4. One of the most crowded regions of the radio signature of the earth: calculated spectra of 0.1-Hz-wide video carriers for channel 3+ (61.260 MHz) as observed from the star Kruger 60 at a distance of 12.9 light years. The four Greenwich sidereal times cover the period during which the star scrapes the northern horizon at lower culmination as seen from the United States. The illustrated frequency offsets arise solely from the Doppler shift due to the rotation of the earth; in reality the degree of crowding is much less since the frequencies of video carriers often wander by as much as ±200 Hz from their assigned frequencies. Individual stations are labelled by numbers.

manage to escape by bouncing off the ground. (Only a negligible portion ever reaches any TV set.) Since most of the power is broadcast near the horizon, only when a star is rising or setting, i.e., is on the horizon as seen from a given antenna location, will it be illuminated with radio power (Figure 3).

After his initial discovery of these radio waves from the direction of our Sun, our eavesdropper would undoubtedly first ask, "Is this some kind of strange natural radio emission, or has some form of civilization produced it?" It would seem that the narrow-band nature of the signals would be one of the best clues that the signal is artificial in nature, as no astrophysical process (known to us) can channel comparable amounts of energy into such small frequency intervals. Other clues, such as polarization of the signals, also exist. And yet, who knows? Perhaps the theorists of another planet are clever enough to come up with a substance whose emission spectrum matches that of the observed radio waves! Clever theorists notwithstanding, for this discussion we assume that the signals from Earth would be recognized as ⁻rtificial.

SCIENTIFIC DEDUCTIONS

As shown in Figure 3, when a star is near the horizon and thus illuminated by a particular station, the station must be near the edge of the Earth as seen from the direction of the star. The result is that the Earth has a very "bright" limb when observed with a receiver for television frequencies (40 to 800 MHz), but the great distance to our eavesdropper's radio telescope means that he is unable actually to discern the disk of the Earth. Nevertheless, the Doppler shift of each station due to the rotation of our planet can tell him not only whether the station is on the approaching or receding side of the Earth, but also whether its latitude is near the fast-spinning equator or the more slowly moving polar regions. Furthermore, he could discover a station's longitude from the times of the twice-daily appearance of the signal from each station. Thus he could construct a map (just like Figure 2, but of course without the outlines of the continents) of all detected stations, each located to an accuracy of a few kilometers.

Because of the extremely nonuniform distribution of stations on the Earth, the total number of stations visible at any one time to an outsider will vary with a 24 (sidereal) hour period. The situation as it would be measured from Barnard's Star is shown in Figure 5. The peaks correspond to the times when population centers with concentrations of television transmitters are on the Earth's limb. By combining data on these intensity variations and Doppler shifts in a straightforward fashion, any eavesdropper could deduce his position relative to our equator (i.e., his declination), the radius of the Earth (6,000 km), and the rotational velocity at the equator (0.5 km/sec).

With this information in hand, the observer is likely to suspect that he is dealing with a planet-like body. His next step might be to study the Earth's annual motion about the Sun (at a rate of 30 km/sec), which causes very large Doppler shifts in the signals of all the individual stations. By tracking these shifts over a year or more, the Earth-Sun system can then be investigated exactly as astronomers study single-line spectroscopic binaries. In such a system two bodies (usually two stars) are orbiting about each other, but only the Doppler shifts in the spectral lines of one member (usually the brighter of the two) can be measured. In the present case the "spectral lines" are the TV carrier signals and the "bright" member is the Earth, far outshining the Sun at the radio frequencies we are discussing. It can be shown that radio observations of the Earth, together with standard optical observations of the associated G2 dwarf star, which we call our Sun, would yield all the vital orbital data for the Earth: its orbital period, its eccentricity, the true Sun-Earth mean distance, etc. The eavesdropper would then be able to provide his colleagues in the Exobiology Department with a good estimate for the Earth's surface temperature, allowing them to place constraints on the possible forms of life responsible for the radio signals. Furthermore, the duration of each station's daily appearances indicates the beam size, and, assuming a diffraction-limited aperture, the dimensions of the transmitting antennas (typically 15 to 20 meters) can be readily deduced, yielding vital clues to the size scales of terrestrial engineering.

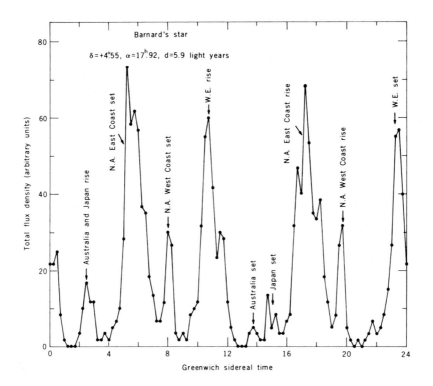

Figure 5. Calculated flux density (summed over all frequencies) of television radiation that would be measured over a sidereal day by an observer located in the direction of Barnard's star, third closest to the Sun and near our celestial equator. The origin of the various peaks is indicated; "rise" and "set" refer respectively to the appearance at the western limb and disappearance at the eastern limb of a particular region on the rotating earth.
Abbreviations: N.A., North America; W.E., Western Europe.

     There are also more subtle effects contained in the TV carrier signals, effects which may or may not remain ambiguous to the observer. For instance, seasonal variations in vegetation, weather, and the iono-sphere will leave their mark in each station's signal. Vegetation has an influence on the amount of power reflected from the surface, as does the choppiness of the sea for coastal stations. The weather and iono-sphere affect the direction and intensity of the radiated power, either through winds flexing the antenna structure or through our upper atmos-phere bending and absorbing the radio waves on their way out. These conditions will cause the observed power levels and times of station appearance to vary slightly and, at first, inexplicably from those predicted. Detailed study may nevertheless allow a few basic conclusions; for example, the presence of an ionized gas around the planet might be deduced from the clue that the lowest frequency stations are much more affected than those at higher frequencies.

A second type of complexity results from such things as a station's daily sign-off hour and the specific frequency and antenna conventions which it follows. These generally vary from one country to another, but can be the same even for countries which are widely separated, but otherwise cooperative in trade, politics or technology. For example, frequency assignments and other conventions are very similar in Japan and the United States. We can interpret these diverse patterns with our detailed cultural and historical knowledge, whereas the extraterrestrial probably cannot - unless his social theory is advanced far beyond our own. The overall problem is not unlike that confronting an archaeologist trying to understand an ancient city with a knowledge of only its street plan. It can only be hoped that the many unsolved puzzles would not hinder the eavesdropper from understanding the more regular and straightforward features of the Earth's radio spectrum.

AN OBSERVATIONAL TEST

In order to sample the radio signature of the earth from an external site and thus test whether television broadcasting is in fact the principal component, S. H. Knowles and the author used the Moon as a handy and objective reflector of the Earth's leakage. Using the 305-m Arecibo radio telescope we scanned a wide range of frequencies between 100 and 400 MHz and found that, once local interference was eliminated (using an on-Moon, off-Moon technique), the frequencies of most observed signals could indeed be identified with the video carriers of various nationalities (Figure 6). Not all the countries mentioned in Figure 6 were on the limb of the Earth as seen from the Moon, however, illustrating that some power leaks also from relatively high elevation angles at the transmitters, although of lesser intensity ($\sim$1-3%).

SHOULD WE TRY TO EAVESDROP?

The above discussion is of course relevant to the larger issue of our own attempts at contact with extraterrestrial civilizations. For the one civilization about which we do know something (our own), note that it has sent out virtually no purposeful signals, yet has been leaking radiation for several decades. How typical this situation will be in our own future or at any time for other galactic civilizations is problematical. Cable television may replace the present system of broadcasting antennas, but new forms of radio leakage may just as well appear. For example, even the slightest bit of back-lobe leakage from the giant transmitting antennas for the Solar Power Satellite scheme, as recently proposed in the U.S., would create extremely intense, repeatable, narrow-band microwave signals. It is true that the range of

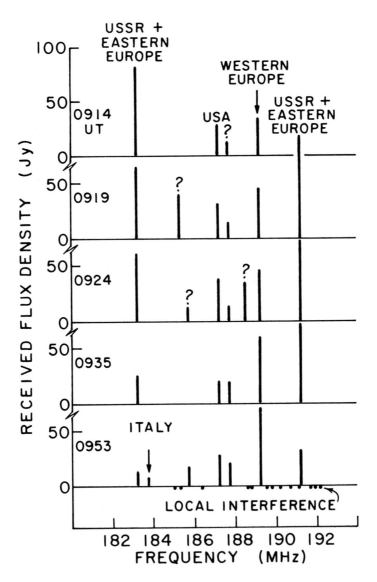

Figure 6. Synoptic series of spectra (10-kHz resolution) in one of the VHF bands assigned to television broadcasting. The Arecibo telescope was pointed alternately on and off the Moon; reflected signals were taken to be only those absent in the off spectrum. Probable identifications, based on international frequency allocations for video carriers, are indicated.

detection of any purposeful beacons is <u>probably</u> much larger than for leaking signals, but another civilization may be leaking prodigious amounts compared to us, or they may have for instance set up powerful navigational beacons for interstellar travel. Furthermore, purposeful signals require decoding of any received message, while we have seen that unintentional signals, so long as they are narrow-band and periodic, yield a great deal of information using only standard astronomical techniques. Not only that, but in a sense the information gained may be a more accurate reflection of the society's major concerns. At least this seems to be true for the case of our own civilization with its military and television leakage, although we might not wish to admit it.

In terms of actual search strategies at the radio telescope, the eavesdropping hypothesis does not come into direct conflict with that of purposeful signals, but rather it suggests that a well-designed system must allow for more possibilities. In particular extremely narrow-band signals drifting in frequency and variable in intensity must not be automatically filtered out. And of course the rationale behind special "natural" frequencies fails, but on the other hand leakage will occur <u>a priori</u> at such frequencies as well as at any others, so they should still be used, although perhaps not exclusively.

In summary then, we should keep both possibilities in mind when searching for extraterrestrial signals. We cannot know whether the most likely signals to be detected would place us in the role of the intended recipient or of eavesdropper.

REFERENCES

Oliver, B.M., and Billingham, J.: *Project Cyclops: A Design Study of a System for Detecting Extraterrestrial Intelligent Life,* NASA Contract Rept. 114445 (Ames Research Center, California, 1973).

Shklovskii, I.S., and Sagan, C., eds.: *Intelligent Life in the Universe* (Delta, New York, 1966), 255-257.

Sullivan, III, W.T., Brown, S., and Wetherill, C., 1978, *Science,* 199, 377; related correspondence in *Science,* 202, 374 ff.

Webb, J.A., in *Interstellar Communication* (A.G.W. Cameron, ed., Benjamin, New York, 1963), 178.

CONCLUSIONS

CONCLUSIONS AND RECOMMENDATIONS FROM THE JOINT SESSION
ON STRATEGIES FOR THE SEARCH FOR LIFE IN THE UNIVERSE

Michael D. Papagiannis
Department of Astronomy, Boston University
Boston, Massachusetts 02215, U.S.A.

The Joint Session in Montreal did not attempt to draw any conclusions, or to draft any recommendations. There was no time for such an effort in a crowded, one day program. The following day, however, in a special open evening session, Frank Drake and myself summarized the results of the meeting for the general membership of the I.A.U. and we made a first attempt to draw some general conclusions and point toward advisable strategies. This is a more formal, though by necessity somewhat subjective attempt to do the same, using not only the oral presentations and discussions from Montreal, but also the more detailed written expositions included in this Volume.

The papers of Part I show that there is a clear dichotomy of opinion on the ability of advanced civilizations to colonize the galaxy. In the early days of SETI 10-20 years ago, it was generally believed that interstellar travel was impossible. A growing number of scientists, however, have espoused in recent years the idea that interstellar travel and hence the colonization of the galaxy is not only feasible but is also inevitable. In a sense this basic disagreement represents a healthy development, because if the entire galaxy is already teeming with intelligent life, it should not be too difficult to find out. Most of our nearby stars and possibly our own solar system would have to be populated with galactic colonies. The obvious strategy, therefore, would be to start our search with a methodical investigation of our immediate stellar neighborhood, which happens also to be the simplest and most practical approach.

The papers of Part II show that radio searches so far have studied less than 1,000 stars and less than 10 galaxies. The searches have been performed primarily at 21 cm, though a few other distinct wavelengths have also been used. All of the results up to now have been negative, but we have only covered an infinitessimally small fraction of the parameter space possible. The development of a near-optimum, low-noise amplifier and of a new generation of multi-channel spectrum analyzers with 10^6-10^7 channels, both of which are now in progress, will improve our bandwidth coverage and frequency resolution by several

243

M. D. Papagiannis (ed.), Strategies for the Search for Life in the Universe, 243–245.
Copyright © 1980 by D. Reidel Publishing Company.

orders of magnitude. It appears, therefore, that with moderate funding
and with already existing radio telescopes we will be able in the decade of the eighties to expand substantially our radio search capabilities.

The papers of Part III show that there is great progress in our
efforts to detect planets in other solar systems. Several new methods
are being tested and several others are being proposed, while the development of an electronic Multichannel Astrometric Photometer (MAP)
promises to increase the present accuracy of astrometric observations
by a factor of 100. The launching of the space telescope during the
80's will be another step in the right direction. It is estimated that
a specially designed space telescope equipped with an MAP will be able
to detect earth-like planets in hundreds of stars. If properly supported, therefore, these studies will increase immensely our knowledge
on the types of planets that orbit around our nearest stars, pointing
out in the process the most promising cases for a thorough search for
extraterrestrial intelligence.

The papers of Part IV indicate that within a few decades our
technology might be able to produce large spaceships capable of interstellar missions at $V \simeq 0.1c$. Of course there is always the problem
of the colossal costs involved in such projects, but the equally colossal military expenditures of practically all nations indicate that
the resources are basically available, but albeit misdirected. Of a
number of suggestions to look for different potentially detectable
manifestations of advanced civilizations, such as Dyson Spheres, etc.,
the proposal to search for unintentional radio signals leaking out from
other civilizations seems to be the most practical one, which will become even more realistic with the use in a few years of the mega-
channel spectrum analyzers.

In conclusion, it appears that our technology has reached a stage
that justifies the investment of time, effort and money in a systematic search for extraterrestrial intelligence. Of course, there is always the question of priorities and limited budgets. It is difficult,
however, to think of a more profound scientific goal than the search
for life in the Universe, and therefore a reasonable, well planned
SETI program should by its nature be given a high priority among the
scientific objectives of our society. This would have to be a long-
range program, which has all the qualifications to develop into an
international undertaking with different nations sharing the responsi-
bilities for different aspects of the program.

In the first phase of the program the emphasis should be placed
on radio searches in the microwave region and in the detection of
planets with astrometric techniques, with parallel development of the
necessary instrumentation. The program should focus initially on our
solar system and 10-20 of the most promising of our neighboring stars.
A systematic microwave search with high spectral resolution over a
wide frequency range should look not only for purposeful signals, but

also for strong radio signals leaking out unintentionally from neighboring civilizations. At the same time, astrometric observations will
concentrate on the same small number of stars in an effort to identify
planets around them. A parallel search should also be mounted in our
solar system to examine asteroids and other minor bodies for anomalous
physical properties that might indicate artificial characteristics.
This search would be carried out with astronomical techniques, supple-
mented whenever possible with observations from space vehicles.

This program can easily be blended into our on going research
programs in Astronomy and Space Sciences, thus keeping the expenses
required for this search to a reasonable level, probably to no more
than a few cents for every 1,000 dollars we spend on defense. The
program should also encourage and support the development of new ideas
which some day might supplement, or even replace our present techniques. Some all-sky surveys could also be undertaken to look for
strong radio beacons operated by far away civilizations, though their
existence seems less probable. In time, the program might expand to
also include sub-millimeter and infrared searches from space vehicles
and might increase the number of stars under scrutiny to several hundred. The first phase of the program, however, ought to be able in
10-15 years to give us a clear indication of whether our galaxy has
already been colonized or not.

If such a search program is pursued dilligently, it is reasonable
to expect that by the turn of the century we will have some impressive
results. It is essential though to realize that negative results are
also important and do not constitute a failure of the search program.
Exciting as it might be to find and communicate with other galactic
civilizations, it would be equally important to know whether we are
one of the precious few, if not the only advanced civilization in our
galaxy. The challenge for generations to come would be enormous, because we could become the starfarers to infuse intelligent life into
the whole galaxy.

We live in a unique period in the history of mankind. Science
and technology have finally made it possible for us to search for other
stellar civilizations, a dream that man has nourished for thousands of
years. Let us proceed.

INDEX

ASTROPHYSICS AND SPACE SCIENCE LIBRARY

Edited by

J. E. Blamont, R. L. F. Boyd, L. Goldberg, C. de Jager, Z. Kopal, G. H. Ludwig, R. Lüst,
B. M. McCormac, H. E. Newell, L. I. Sedov, Z. Švestka, and W. de Graaff

1. C. de Jager (ed.), *The Solar Spectrum, Proceedings of the Symposium held at the University of Utrecht, 26–31 August, 1963.* 1965, XIV + 417 pp.
2. J. Orthner and H. Maseland (eds.), *Introduction to Solar Terrestrial Relations, Proceedings of the Summer School in Space Physics held in Alpbach, Austria, July 15–August 10, 1963 and Organized by the European Preparatory Commission for Space Research.* 1965, IX + 506 pp.
3. C. C. Chang and S. S. Huang (eds.), *Proceedings of the Plasma Space Science Symposium, held at the Catholic University of America, Washington, D.C., June 11–14, 1963.* 1965, IX + 377 pp.
4. Zdeněk Kopal, *An Introduction to the Study of the Moon.* 1966, XII + 464 pp.
5. B. M. McCormac (ed.), *Radiation Trapped in the Earth's Magnetic Field. Proceedings of the Advanced Study Institute, held at the Chr. Michelsen Institute, Bergen, Norway, August 16– September 3, 1965.* 1966, XII + 901 pp.
6. A. B. Underhill, *The Early Type Stars.* 1966, XII + 282 pp.
7. Jean Kovalevsky, *Introduction to Celestial Mechanics.* 1967, VIII + 427 pp.
8. Zdeněk Kopal and Constantine L. Goudas (eds.), *Measure of the Moon. Proceedings of the 2nd International Conference on Selenodesy and Lunar Topography, held in the University of Manchester, England, May 30–June 4, 1966.* 1967, XVIII + 479 pp.
9. J. G. Emming (ed.), *Electromagnetic Radiation in Space. Proceedings of the 3rd ESRO Summer School in Space Physics, held in Alpbach, Austria, from 19 July to 13 August, 1965.* 1968, VIII + 307 pp.
10. R. L. Carovillano, John F. McClay, and Henry R. Radoski (eds.), *Physics of the Magnetosphere, Based upon the Proceedings of the Conference held at Boston College, June 19–28, 1967.* 1968, X + 686 pp.
11. Syun-Ichi Akasofu, *Polar and Magnetospheric Substorms.* 1968, XVIII + 280 pp.
12. Peter M. Millman (ed.), *Meteorite Research. Proceedings of a Symposium on Meteorite Research, held in Vienna, Austria, 7–13 August, 1968.* 1969, XV + 941 pp.
13. Margherita Hack (ed.), *Mass Loss from Stars. Proceedings of the 2nd Trieste Colloquium on Astrophysics, 12–17 September, 1968.* 1969, XII + 345 pp.
14. N. D'Angelo (ed.), *Low-Frequency Waves and Irregularities in the Ionosphere. Proceedings of the 2nd ESRIN-ESLAB Symposium, held in Frascati, Italy, 23–27 September, 1968.* 1969, VII + 218 pp.
15. G. A. Partel (ed.), *Space Engineering. Proceedings of the 2nd International Conference on Space Engineering, held at the Fondazione Giorgio Cini, Isola di San Giorgio, Venice, Italy, May 7–10, 1969.* 1970, XI + 728 pp.
16. S. Fred Singer (ed.), *Manned Laboratories in Space. Second International Orbital Laboratory Symposium.* 1969, XIII + 133 pp.
17. B. M. McCormac (ed.), *Particles and Fields in the Magnetosphere. Symposium Organized by the Summer Advanced Study Institute, held at the University of California, Santa Barbara, Calif., August 4–15, 1969.* 1970, XI + 450 pp.
18. Jean-Claude Pecker, *Experimental Astronomy.* 1970, X + 105 pp.
19. V. Manno and D. E. Page (eds.), *Intercorrelated Satellite Observations related to Solar Events. Proceedings of the 3rd ESLAB/ESRIN Symposium held in Noordwijk, The Netherlands, September 16–19, 1969.* 1970, XVI + 627 pp.
20. L. Mansinha, D. E. Smylie, and A. E. Beck, *Earthquake Displacement Fields and the Rotation of the Earth, A NATO Advanced Study Institute Conference Organized by the Department of Geophysics, University of Western Ontario, London, Canada, June 22–28, 1969.* 1970, XI + 308 pp.
21. Jean-Claude Pecker, *Space Observatories.* 1970, XI + 120 pp.
22. L. N. Mavridis (ed.), *Structure and Evolution of the Galaxy. Proceedings of the NATO Advanced Study Institute, held in Athens, September 8–19, 1969.* 1971, VII + 312 pp.

23. A. Muller (ed.), *The Magellanic Clouds. A European Southern Observatory Presentation: Principal Prospects, Current Observational and Theoretical Approaches, and Prospects for Future Research, Based on the Symposium on the Magellanic Clouds, held in Santiago de Chile, March 1969, on the Occasion of the Dedication of the European Southern Observatory.* 1971, XII + 189 pp.

24. B. M. McCormac (ed.), *The Radiating Atmosphere. Proceedings of a Symposium Organized by the Summer Advanced Study Institute, held at Queen's University, Kingston, Ontario, August 3–14, 1970.* 1971, XI + 455 pp.

25. G. Fiocco (ed.), *Mesospheric Models and Related Experiments. Proceedings of the 4th ESRIN-ESLAB Symposium, held at Frascati, Italy, July 6–10, 1970.* 1971, VIII + 298 pp.

26. I. Atanasijević, *Selected Exercises in Galactic Astronomy.* 1971, XII + 144 pp.

27. C. J. Macris (ed.), *Physics of the Solar Corona. Proceedings of the NATO Advanced Study Institute on Physics of the Solar Corona, held at Cavouri-Vouliagmeni, Athens, Greece, 6–17 September 1970.* 1971, XII + 345 pp.

28. F. Delobeau, *The Environment of the Earth.* 1971, IX + 113 pp.

29. E. R. Dyer (general ed.), *Solar-Terrestrial Physics/1970. Proceedings of the International Symposium on Solar-Terrestrial Physics, held in Leningrad, U.S.S.R., 12–19 May 1970.* 1972, VIII + 938 pp.

30. V. Manno and J. Ring (eds.), *Infrared Detection Techniques for Space Research. Proceedings of the 5th ESLAB-ESRIN Symposium, held in Noordwijk, The Netherlands, June 8–11, 1971.* 1972, XII + 344 pp.

31. M. Lecar (ed.), *Gravitational N-Body Problem. Proceedings of IAU Colloquium No. 10, held in Cambridge, England, August 12–15, 1970.* 1972, XI + 441 pp.

32. B. M. McCormac (ed.), *Earth's Magnetospheric Processes. Proceedings of a Symposium Organized by the Summer Advanced Study Institute and Ninth ESRO Summer School, held in Cortina, Italy, August 30–September 10, 1971.* 1972, VIII + 417 pp.

33. Antonin Rükl, *Maps of Lunar Hemispheres.* 1972, V + 24 pp.

34. V. Kourganoff, *Introduction to the Physics of Stellar Interiors.* 1973, XI + 115 pp.

35. B. M. McCormac (ed.), *Physics and Chemistry of Upper Atmospheres. Proceedings of a Symposium Organized by the Summer Advanced Study Institute, held at the University of Orléans, France, July 31–August 11, 1972.* 1973, VIII + 389 pp.

36. J. D. Fernie (ed.), *Variable Stars in Globular Clusters and in Related Systems. Proceedings of the IAU Colloquium No. 21, held at the University of Toronto, Toronto, Canada, August 29–31, 1972.* 1973, IX + 234 pp.

37. R. J. L. Grard (ed.), *Photon and Particle Interaction with Surfaces in Space. Proceedings of the 6th ESLAB Symposium, held at Noordwijk, The Netherlands, 26–29 September, 1972.* 1973, XV + 577 pp.

38. Werner Israel (ed.), *Relativity, Astrophysics and Cosmology. Proceedings of the Summer School, held 14–26 August, 1972, at the BANFF Centre, BANFF, Alberta, Canada.* 1973, IX + 323 pp.

39. B. D. Tapley and V. Szebehely (eds.), *Recent Advances in Dynamical Astronomy. Proceedings of the NATO Advanced Study Institute in Dynamical Astronomy, held in Cortina d'Ampezzo, Italy, August 9–12, 1972.* 1973, XIII + 468 pp.

40. A. G. W. Cameron (ed.), *Cosmochemistry. Proceedings of the Symposium on Cosmochemistry, held at the Smithsonian Astrophysical Observatory, Cambridge, Mass., August 14–16, 1972.* 1973, X + 173 pp.

41. M. Golay, *Introduction to Astronomical Photometry.* 1974, IX + 364 pp.

42. D. E. Page (ed.), *Correlated Interplanetary and Magnetospheric Observations. Proceedings of the 7th ESLAB Symposium, held at Saulgau, W. Germany, 22–25 May, 1973.* 1974, XIV + 662 pp.

43. Riccardo Giacconi and Herbert Gursky (eds.), *X-Ray Astronomy.* 1974, X + 450 pp.

44. B. M. McCormac (ed.), *Magnetospheric Physics. Proceedings of the Advanced Summer Institute, held in Sheffield, U.K., August 1973.* 1974, VII + 399 pp.

45. C. B. Cosmovici (ed.), *Supernovae and Supernova Remnants. Proceedings of the International Conference on Supernovae, held in Lecce, Italy, May 7–11, 1973.* 1974, XVII + 387 pp.

46. A. P. Mitra, *Ionospheric Effects of Solar Flares.* 1974, XI + 294 pp.

47. S.-I. Akasofu, *Physics of Magnetospheric Substorms.* 1977, XVIII + 599 pp.

48. H. Gursky and R. Ruffini (eds.), *Neutron Stars, Black Holes and Binary X-Ray Sources*. 1975, XII + 441 pp.

49. Z. Švestka and P. Simon (eds.), *Catalog of Solar Particle Events 1955–1969. Prepared under the Auspices of Working Group 2 of the Inter-Union Commission on Solar-Terrestrial Physics*. 1975, IX + 428 pp.

50. Zdeněk Kopal and Robert W. Carder, *Mapping of the Moon*. 1974, VIII + 237 pp.

51. B. M. McCormac (ed.), *Atmospheres of Earth and the Planets. Proceedings of the Summer Advanced Study Institute, held at the University of Liège, Belgium, July 29–August 8, 1974*. 1975, VII + 454 pp.

52. V. Formisano (ed.), *The Magnetospheres of the Earth and Jupiter. Proceedings of the Neil Brice Memorial Symposium, held in Frascati, May 28–June 1, 1974*. 1975, XI + 485 pp.

53. R. Grant Athay, *The Solar Chromosphere and Corona: Quiet Sun*. 1976, XI + 504 pp.

54. C. de Jager and H. Nieuwenhuijzen (eds.), *Image Processing Techniques in Astronomy. Proceedings of a Conference, held in Utrecht on March 25–27, 1975*. XI + 418 pp.

55. N. C. Wickramasinghe and D. J. Morgan (eds.), *Solid State Astrophysics. Proceedings of a Symposium, held at the University College, Cardiff, Wales, 9–12 July 1974*. 1976, XII + 314 pp.

56. John Meaburn, *Detection and Spectrometry of Faint Light*. 1976, IX + 270 pp.

57. K. Knott and B. Battrick (eds.), *The Scientific Satellite Programme during the International Magnetospheric Study. Proceedings of the 10th ESLAB Symposium, held at Vienna, Austria, 10–13 June 1975*. 1976, XV + 464 pp.

58. B. M. McCormac (ed.), *Magnetospheric Particles and Fields. Proceedings of the Summer Advanced Study School, held in Graz, Austria, August 4–15, 1975*. 1976, VII + 331 pp.

59. B. S. P. Shen and M. Merker (eds.), *Spallation Nuclear Reactions and Their Applications*. 1976, VIII + 235 pp.

60. Walter S. Fitch (ed.), *Multiple Periodic Variable Stars. Proceedings of the International Astronomical Union Colloquium No. 29, held at Budapest, Hungary, 1–5 September 1976*. 1976, XIV + 348 pp.

61. J. J. Burger, A. Pedersen, and B. Battrick (eds.), *Atmospheric Physics from Spacelab. Proceedings of the 11th ESLAB Symposium, Organized by the Space Science Department of the European Space Agency, held at Frascati, Italy, 11–14 May 1976*. 1976, XX + 409 pp.

62. J. Derral Mulholland (ed.), *Scientific Applications of Lunar Laser Ranging. Proceedings of a Symposium held in Austin, Tex., U.S.A., 8–10 June, 1976*. 1977, XVII + 302 pp.

63. Giovanni G. Fazio (ed.), *Infrared and Submillimeter Astronomy. Proceedings of a Symposium held in Philadelphia, Penn., U.S.A., 8–10 June, 1976*. 1977, X + 226 pp.

64. C. Jaschek and G. A. Wilkins (eds.), *Compilation, Critical Evaluation and Distribution of Stellar Data. Proceedings of the International Astronomical Union Colloquium No. 35, held at Strasbourg, France, 19–21 August, 1976*. 1977, XIV + 316 pp.

65. M. Friedjung (ed.), *Novae and Related Stars. Proceedings of an International Conference held by the Institut d'Astrophysique, Paris, France, 7–9 September, 1976*. 1977, XIV + 228 pp.

66. David N. Schramm (ed.), *Supernovae. Proceedings of a Special IAU-Session on Supernovae held in Grenoble, France, 1 September, 1976*. 1977, X + 192 pp.

67. Jean Audouze (ed.), *CNO Isotopes in Astrophysics. Proceedings of a Special IAU Session held in Grenoble, France, 30 August, 1976*. 1977, XIII + 195 pp.

68. Z. Kopal, *Dynamics of Close Binary Systems*, XIII + 510 pp.

69. A. Bruzek and C. J. Durrant (eds.), *Illustrated Glossary for Solar and Solar-Terrestrial Physics*. 1977, XVIII + 204 pp.

70. H. van Woerden (ed.), *Topics in Interstellar Matter*. 1977, VIII + 295 pp.

71. M. A. Shea, D. F. Smart, and T. S. Wu (eds.), *Study of Travelling Interplanetary Phenomena*. 1977, XII + 439 pp.

72. V. Szebehely (ed.), *Dynamics of Planets and Satellites and Theories of Their Motion. Proceedings of IAU Colloquium No. 41, held in Cambridge, England, 17–19 August 1976*. 1978, XII + 375 pp.

73. James R. Wertz (ed.), *Spacecraft Attitude Determination and Control*. 1978, XVI + 858 pp.

74. Peter J. Palmadesso and K. Papadopoulos (eds.), *Wave Instabilities in Space Plasmas. Proceedings of a Symposium Organized Within the XIX URSI General Assembly held in Helsinki, Finland, July 31–August 8, 1978.* 1979, VII + 309 pp.

75. Bengt E. Westerlund (ed.), *Stars and Star Systems. Proceedings of the Fourth European Regional Meeting in Astronomy held in Uppsala, Sweden, 7–12 August, 1978.* 1979, XVIII + 264 pp.

76. Cornelis van Schooneveld (ed.), *Image Formation from Coherence Functions in Astronomy. Proceedings of IAU Colloquium No. 49 on the Formation of Images from Spatial Coherence Functions in Astronomy, held at Groningen, The Netherlands, 10–12 August 1978.* 1979, XII + 338 pp.

77. Zdeněk Kopal, *Language of the Stars. A Discourse on the Theory of the Light Changes of Eclipsing Variables.* 1979, VIII + 280 pp.

78. S.-I. Akasofu (ed.), *Dynamics of the Magnetosphere. Proceedings of the A.G.U. Chapman Conference 'Magnetospheric Substorms and Related Plasma Processes' held at Los Alamos Scientific Laboratory, N.M., U.S.A., October 9–13, 1978.* 1980, XII + 658 pp.

79. Paul S. Wesson, *Gravity, Particles, and Astrophysics. A Review of Modern Theories of Gravity and G-variability, and their Relation to Elementary Particle Physics and Astrophysics.* 1980, VIII + 188 pp.

80. Peter A. Shaver (ed.), *Radio Recombination Lines. Proceedings of a Workshop held in Ottawa, Ontario, Canada, August 24-25, 1979.* 1980, X + 284 pp.

81. Pier Luigi Bernacca and Remo Ruffini (eds.), *Astrophysics from Spacelab*, 1980, XI + 664 pp.